The
Board Designer's Guide
to
Testable Logic Circuits

ELECTRONIC SYSTEMS ENGINEERING SERIES

Consulting editors   **E L Dagless**
University of Bristol

J O'Reilly
University College of Wales

OTHER TITLES IN THE SERIES

Advanced Microprocessor Architectures   *L Ciminiera and A Valenzano*

Optical Pattern Recognition Using Holographic Techniques   *N Collings*

Modern Logic Design   *D Green*

Data Communications, Computer Networks and OSI (2nd Edn)   *F Halsall*

Multivariable Feedback Design   *J M Maciejowski*

Microwave Components and Systems   *K F Sander*

Tolerance Design of Electronic Circuits   *R Spence and R Soin*

Computer Architecture and Design   *A J van de Goor*

Digital Systems Design with Programmable Logic   *M Bolton*

Introduction to Robotics   *P J McKerrow*

MAP and TOP Communications: Standards and Applications   *A Valenzano, C Demartini and L Ciminiera*

Integrated Broadband Networks   *R Händel and M N Huber*

The
Board Designer's Guide
to
Testable Logic Circuits

COLIN MAUNDER
B T Laboratories, Martlesham Heath, Ipswich

Addison-Wesley Publishing Company

Wokingham, England • Reading, Massachusetts • Menlo Park, California • New York
Don Mills, Ontario • Amsterdam • Bonn • Sydney • Singapore • Tokyo • Madrid
San Juan • Milan • Paris • Mexico City • Seoul • Taipei

Cover designed by Designers & Partners of Oxford
and printed by The Riverside Printing Co. (Reading) Ltd.
Printed in Great Britain at the University Press, Cambridge.

First printed 1991.

British Library Cataloguing in Publication Data
Maunder, Colin
   The board users guide to testable logic circuits.
   I. Title
   621.38153

   ISBN 0–201–56513–7

Library of Congress Cataloging-in-Publication Data
Maunder, Colin M.
   The board designers guide to testable logic circuits / by Colin Maunder.
      p.   cm. — (Electronic systems engineering series)
   Includes bibliographical references and index.
   ISBN 0–201–56513–7
   1. Logic circuits—Testing.   I. Title.   II. Series.
TK7868.L6M376   1992
621.39'5—dc20   91–29663
                                           CIP

Preface

When I first got involved in test engineering in the 1970s, boards were, by today's standards, simple. They contained, on average, a few thousand logic gates in the form of small- and medium-scale integration components. Testing was done by applying signals at the board's functional connector and by examining the board's response.

Of course, board complexities increased. Soon, it was recognized that functional (from the edge) testing was too costly — it was expensive to develop test programs and these were relatively inefficient at locating common manufacturing faults, such as open and short circuits. The in-circuit tester was introduced as a solution to these problems. It allowed test generation costs to be significantly reduced and, by virtue of its connection to every chip-to-chip interconnection on the board, allowed rapid diagnosis of the most common manufacturing-induced faults.

During the 1980s, the in-circuit tester became the principal type of test system used for testing loaded boards. Initially, some were concerned that the backdriving technique used by these testers might cause damage or reliability degradation to components on the boards. In response, techniques were developed to control the way that in-circuit tests were applied — ensuring that the chance of damage was minimized. In essence, these techniques required careful sequencing of the tests applied to the board, allowing a recovery period following backdriving of a particular IC, and imposition of a maximum time limit for each test, calculated according to the characteristics of the components adjacent to that under test.

Unfortunately, technology doesn't stand still. Board complexities continued to increase during the 1980s, fuelled by advancing integrated circuit technology and by the move towards the use of smaller surface-mount packages. This caused three problems for the in-circuit tester:

❐ First, test times for individual components increased and began to exceed the time limit imposed to avoid the possibility of damage during backdriving. Tests had to be shortened, with the result that they were less comprehensive than before.

❐ Second, the pin-to-pin spacing for surface-mount packages is less than the 0.1" of dual-in-line ICs. The spacing between in-circuit test probes had to be reduced to allow connections into and out of these ICs to be accessed. Unfortunately, however, probes become less robust and less reliable as their size reduces.

Towards the end of the 1980s, these problems were becoming acute in certain sectors of the electronics industry. An industry pressure group (the Joint Test Action Group — JTAG) was formed to develop and promulgate a change of approach — from in-circuit testing to a technique more suited to highly-complex, miniaturized loaded board designs. JTAG and, subsequently, the IEEE drafted a standard for the design of integrated circuits that would ensure that chips would be able to assist in the task of testing the loaded board. This standard — *ANSI/IEEE Std 1149.1, Standard Test Access Port and Boundary-Scan Architecture* — is now supported by several leading IC vendors and test equipment companies and is set to provide the basis of board testing through the 1990s.

While these changes have helped to control the cost of testing (which would otherwise have risen much more rapidly as board complexity increased), it is clear that testing has become an expensive part of the total cost of developing, manufacturing, and (in particular) supporting electronic systems. As a result, design-for-test has become an essential aspect of the designer's task — those who ignore it do so at their peril.

The objective of this book is to present design-for-test in a manner that matches the way that a board is designed — starting with top-level block design and progressing through component selection and circuit design to board layout. The design-for-test requirements that should be considered at each stage are, wherever possible, grouped into a single chapter. An explanation is provided for each requirement so that the designer can understand why the feature is needed and what the consequences of ignoring the requirement might be. Finally, a set of checklists is provided to help assess the testability of each completed design — again, stage by stage.

As you will see, design-for-test is not difficult. The various requirements are easy to understand and to implement. If implemented, the various design-for-test features will significantly reduce the cost of testing the finished board design. So why not give it a try?

Acknowledgements

Many people have contributed to this book. Particular acknowledgement should be given to the following:

❐ my employer, BT, for providing the time to develop the internal design-for-test 'manual' that was the basis of this book and for giving permission for its publication;

❐ Ben Bennetts, of Bennetts Associates, who introduced me to test engineering and with whom I have lectured on test generation and design-for-test over many years;

❐ the students on courses presented in Amsterdam and elsewhere, who unwittingly acted as the test-bed for the material in this book and, in some cases, contributed items that have been included in it;

❐ Ken Totton and other colleagues at BT Laboratories, who provided comments on early drafts; and, lastly,

❐ my wife and children, who were starved of attention at evenings and weekends as the draft was converted into this book.

Finally, Chapter 3 contains material from two previously-published papers:

❐ Maunder C.M. and Tulloss R.E. (1991). An introduction to the boundary-scan standard: ANSI/IEEE Std 1149.1. *Journal of Electronics Test, Theory, and Applications,* **2**(1), 27-42.

❐ Maunder C.M. (1991). The impact of testability standards on design and EDA. In *Proc. Concurrent Engineering and Electronic Design Automation Conference.* Bournemouth, UK, March.

Acknowledgement is made to Kluwer Academic Publishers and Bournemouth Polytechnic, respectively, for permission to reuse this material.

Colin Maunder
BT Laboratories
Martlesham Heath
Ipswich, IP5 7RE, UK

November 11, 1991

Contents

Glossary

AC	Alternating current
ACL	Approved component list
ALU	Arithmetic logic unit
AOT	Adjust on test
ASIC	Application-specific IC
ATE	Automatic test equipment
ATPG	Automatic test program generation
BILBO	Built-in logic block observer
BIST	Built-in self-test
BOM	Bill of materials
BSDL	Boundary-scan description language
CCITT	International Telephone and Telegraph Consultative Committee
CMOS	Complementary metal oxide semiconductor
DC	Direct current
DIL	Dual-in-line
DIP	Dual-in-line package
DTL	Diode transistor logic
DUT	Device-under-test
ECL	Emitter-coupled logic
EDA	Electronic design automation
EMC	Electromagnetic compatibility
FBT	Functional board tester: a type of ATE

FDDI	Fibre distributed data interface
IC	Integrated circuit
ICT	In-circuit tester: a type of ATE
ISDN	Integrated services digital network
JEDEC	Joint Electron Device Engineering Council
JTAG	Joint Test Action Group
LED	Light-emitting diode
LFSR	Linear-feedback shift register
LSSD	Level-sensitive scan design
MISR	Multiple-input signature register
MOS	Metal oxide semiconductor
Overdrive	Force the logic state of a connection by supplying more current than the circuit that would normally drive it
PBX	Private branch exchange: a telephone and/or data exchange installed on a company's premises, etc.
PCB	Printed circuit board
PLD	Programmable logic device
PTH	Plated through hole
PWB	Printed wiring board
RAM	Random-access memory
ROM	Read-only memory
RTL	Resistor transistor logic
S-a-0	Stuck-at-0: a fault in which a connection becomes permanently fixed at logic 0
S-a-1	Stuck-at-1: a fault in which a connection becomes permanently fixed at logic 1
SIP	Single-in-line package
SMT	Surface-mount technology
SOT	Select on test
Stuck-at fault	A fault that causes a signal to take on a fixed logic value regardless of the output of preceeding logic
TAP	The ANSI/IEEE Std 1149.1 test access port
Test fixture	Hardware used to connect the printed circuit board being tested to the ATE.
Test Land	Contact area on a printed circuit board used by bed-of-nails test probes
TTL	Transistor transistor logic
VHDL	VHSIC hardware description language
VHSIC	Very high speed integrated circuit
Via	A plated-through hole on a printed circuit board used solely to convey a connection from one layer to another
VLSI	Very large scale integration
UUT	Unit under test
ZIF	Zero insertion force

Part 1

This part provides an introduction to testing and design-for-testability. It provides an overview of the principal design-for-test techniques, for both chips and loaded boards, and discusses the formulation of plans to ensure that products are designed to be testable.

CHAPTER 1.
Introduction to Testing and Testability

1.1. Introduction

A combination of two factors — greater competition and the wider use of 'information technology' — is bringing about a rapid growth in the variety of electronic products available to the consumer. Also evident is the increasing complexity of these products, made possible through the combination of low-cost state of the art integrated circuit technology and advanced manufacturing techniques.

To compete successfully in such an environment, companies need to bring new products from the drawing board to the marketplace as quickly (and as cheaply) as possible, and to encourage their suppliers to do the same. The costly and time-consuming 'production engineering' phases that have traditionally followed initial design must now be avoided, with the consequence that the responsibility for production engineering tasks is increasingly placed on the designer. The areas of design covered by these tasks do, however, contribute significantly to the product's commercial viability and it is therefore important that they continue to be considered carefully.

One such design area is 'design-for-testability' which stems from the need to make the process of testing the product, both following production and during repair, as cost-effective as possible. The purpose of this chapter is to explain why design-for-testability (and hence this book) is needed through a discussion of test technology and the problems of test generation, test application and fault diagnosis.

1.2. Basics

1.2.1. Types of testing

Testing is performed at a number of stages in the development of a product and for a variety of purposes. Perhaps the most important of these types of testing are:

(1) *Design Verification Testing.* Carried out to ensure that the design adequately performs the function which is expected of it. This stage of testing is most often performed using bench-top instrumentation (oscilloscopes, logic analysers, etc.), although in the case of more complex designs programmable test systems may be used.

(2) *Production Testing.* Performed to locate any defects which might exist in each copy of a design once it has been manufactured. Typically, this stage of testing will be done using a programmable **automated test equipment** (ATE), which for assembled printed circuit boards may cost $1,000,000 or more. The principal types of ATE are described in Section 1.4.1.

(3) *Repair Testing.* Performed when a product fails during use. The aim is to isolate the cause of failure sufficiently to allow it to be repaired. Once again, this may be accomplished using ATE.

(4) *Self-testing.* An example of self-test is the routining of telephone exchanges, computers, or military equipment during idle periods, performed to locate faults before they cause failure in use. Self-test procedures are also provided for other reasons, for example to reduce the costs of performing on-site repair. They are particularly suitable for use in equipment that is to be installed in a customer's premises (for example, office equipment or telephone switches) since they allow the customer to check which piece of equipment is at fault before asking for repair.

This book is directed at all stages of testing a product during its life and at ways in which products can be designed to make test tasks easier.

1.2.2. Test activities

There are three stages of activity involved in the testing of a product.

(1) *Test Development.* Firstly, a test for the product must be developed, and demonstrated to be sufficiently good at detecting and locating faults. We will look at one typical method for achieving this in Section 1.3.

(2) *Test Application.* Secondly, once the test has been developed, it will be applied to units as they leave the production line, or as they arrive for repair, using the available ATE. After this stage units will have been marked faulty or fault-free. Types of ATE, and other aspects of test application, are considered in Section 1.4.

(3) *Diagnosis.* Finally, if a unit is found faulty, the ATE will be used to produce a diagnosis of the cause of failure, for example a failed component. In this way, a repair can be effected, the success of which will be determined by re-testing the unit. Ways in which automated diagnosis is accomplished are discussed in Section 1.5.

1.2.3. What is testability?

Since cost is an important factor in any commercial environment, it should not be surprising that the overall cost of performing testing activities discussed in Section 1.2.2 is one measure of the product's testability. The higher the cost of each activity, the lower the product's testability.

Cost is, however, not the only factor which determines testability, and for the purpose of this discussion two other factors will be considered.

Firstly, the time taken to perform the various test tasks is important, and cannot always be measured through the costs of labour, etc. In some cases, for example when deadlines have to be met, time may actually be more important than cost. In these cases, the most significant impact of reduced testability may be lengthened test development timescales. Figure 1.1 (Reinerstein, 1983) shows the importance of time-to-market in high-growth markets where product life cycles are relatively short. It can be seen that a loss of up to 33% of profits may occur if a product is six months late on to the market.

Secondly, there is the adverse effect that reduced testability can have on the quality of the product. Ideally, a test program should be able to detect 100% of the faults that might occur in the product; if this were the case, then the quality of the product could be guaranteed. Inevitably, the test will be less than perfect (as will be discussed later) and so some faults will escape

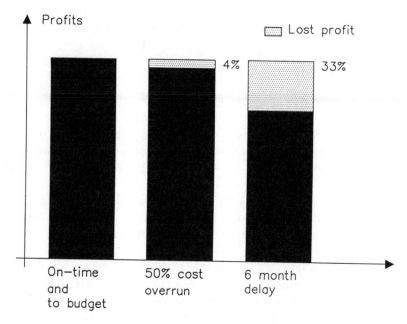

Figure 1.1 Profit loss through exceeded budgets.

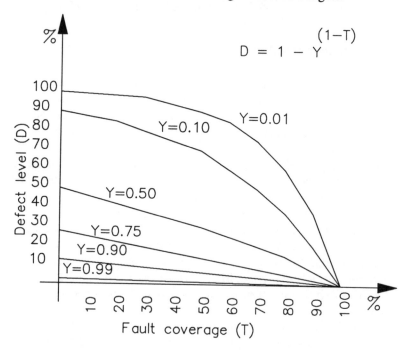

Figure 1.2 Defect level as a function of process yield and fault coverage.

detection, the number depending on the difficulty of testing the product — its 'untestability'.

Figure 1.2 (Williams and Brown, 1981) shows how three factors — process yield (that is, the fraction of products manufactured fault-free), fault coverage (test performance), and shipped defect level (quality) — are related. It is clear that high-performance tests are required if the quality of shipped products is to be acceptable, particularly when the yield of the production process is low.

The possibility that units passed as fault-free by the test might actually contain faults will, of course, appear as a reduction in quality when the product reaches the consumer. Frequently, therefore, *a business's reputation can be damaged if product testability is inadequate.*

1.2.4. Why is testability important?

Figure 1.3 gives some idea of the way in which test costs in the semiconductor industry have risen with each advance in integrated circuit technology.

For a **very-large-scale-integration** (VLSI) integrated circuit, the development of an adequate test program can account for a major part of the overall cost of bringing the chip into production. Chip manufacturers are increasingly resorting to rigid 'design-for-testability' procedures as a means of reducing this expenditure. Indeed, the indications are that it will be impossible for chips as complex as those now starting to come onto the market to be produced economically unless such design procedures are used. To give a rough idea of costs, a VLSI chip may require 24 man-months or more of effort solely in test development — say, at a cost of $200,000.

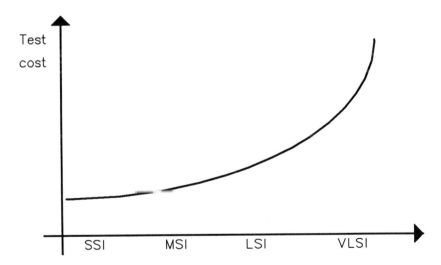

Figure 1.3 Test cost trends.

Obviously, chips of this complexity are a major cause of testability problems in the systems into which they are assembled. Not surprisingly, therefore, the costs of testing for circuit cards and systems are also increasing rapidly.

This raises another point: the need for the designer not only to design his product to be testable in its own right but also to ensure that it does not contribute to poor testability when it is assembled into other products. Testable chips can easily be assembled into untestable systems! As one test engineer put it in 1979:

> *'LSI technology has brought the design of truly untestable circuits within the reach of everyone!'*

A further motivation for high testability comes from the 'Rule of 10s'. This empircal rule, which is widely accepted by the test engineering community, relates the cost of testing for a fault at various stages in product assembly. For example, assume that there is a single fault present in a newly manufactured integrated circuit and that the cost of testing the chip is $C. If the fault is detected by the chip test, the cost is $C.

If the fault is not detected by the chip test, then the faulty integrated circuit (IC) will be assembled into a loaded board. The Rule of 10s predicts that the cost of finding the fault while testing the board and then effecting a repair is $10 x C.

Of course, if the quality of the test program is insufficient (due to poor testability), the fault may not be detected by the loaded board test. In this case, the faulty board will be inserted in a system. The Rule of 10s predicts that the cost of finding the fault during the system test and then effecting a repair will be $100 x C.

Finally, if the fault escapes detection during system test, a faulty system will be shipped to a customer. *Customers always find faults that have not been detected by the manufacturer!* The Rule of 10s predicts that the cost of correcting the fault once the system has been installed in the field will be $1000 x C.

1.3. Test generation

1.3.1. Fault models

Before starting to develop a test for a logic circuit we need to know precisely what the objectives of the test are in terms of the faults it should be able to detect.

Ideally, we would like the test to be able to detect all failures which might occur. However, if each of the failure modes of each transistor, resistor, etc. in the circuit is considered, together with the possibility of unwanted connections between each pair of signals, it becomes apparent that the number of different failures that could potentially occur (many of them are extremely unlikely!) is very large, even for a fairly small circuit. For larger circuits the number of possible failures would very rapidly become prohibitive.

To produce a manageable objective for the test it is common to work in terms of a small range of fault models, each of which covers a range of failure modes. These fault models represent the effects (rather than the causes) of the malfunction and are significantly easier to comprehend. The more common of these fault models are illustrated in Figure 1.4 and described below.

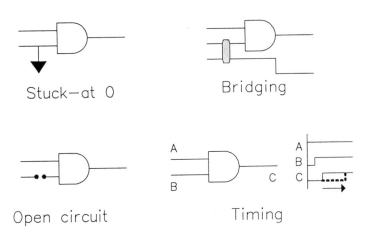

Figure 1.4 Simple fault models.

The 'stuck-at' fault model

The 'stuck-at' fault model was originally proposed in 1959 when the dominant logic technologies were **resistor transistor logic** (RTL) and **diode transistor logic** (DTL) (Eldred, 1959).

In these technologies, almost all failures of circuit components result in one or more circuit connections becoming stuck at one or other of the two logic values (0 or 1) and these failures could therefore be said to result in nodes becoming 'stuck-at 0' (s-a-0) or 'stuck-at 1' (s-a-1).

Despite the fact that integrated circuit technology has advanced considerably 'stuck-at' models are still used widely today. This is because of two factors. Firstly, the models are extremely easy to introduce into circuit simulations — they merely result in nodes being permanently assigned the appropriate logic value. This means that it is relatively easy to ensure that the test is good at detecting stuck-at faults. Secondly, experience has shown that tests which are effective at detecting stuck-at faults are effective at detecting the types of failure more representative of modern technology —

for example failures in **metal oxide semiconductor** (MOS) integrated circuits and **random-access memories** (RAMs).

The 'open circuit' fault model

Open circuit failures in **transistor transistor logic** (TTL) are equivalent to the stuck-at fault models, since most undriven TTL device inputs will 'float' to a logic 1 state. For this reason, open circuit failures may not be considered separately for TTL-compatible circuits.

In MOS technologies, however, the same assumption cannot be made since 'floating' inputs behave in considerably different ways. The use of specific models for open circuit failures during integrated circuit test development is therefore increasing.

The bridging fault model

An extremely common production defect, both for integrated circuits and for circuit boards, is for two nodes to become shorted together. Such a defect is termed a bridging fault, since in circuit board production it is the result of accidental bridging connections being inserted by solder splashes. In integrated circuits the same defect is caused by the failure of insulation between areas of silicon, conductors, and so on. Obviously, many potential bridging faults are unlikely to occur since the relevant connections are too widely spaced. For this reason only specific sets of bridging faults are simulated. Common examples are those between adjacent pins on a device or between adjacent PWB tracks since these are most likely to be shorted by extraneous solder.

Note that it is important to model the result of the short-circuit fault correctly. For some faults, the outcome will be the wire-OR combination of the signals that drive the shorted tracks, while for others the outcome may be the wire-AND or an indeterminate voltage level.

The timing fault model

Fortunately by avoiding the use of asynchronous circuits (which is good design practice anyway) the likelihood of drifts in the timing behaviour of devices causing the failure of a circuit can be made extremely small. However, in circuits where accurate timings are required for successful operation, the possibility of such timing drifts must be considered and this is done using the timing fault model. In this model, the possibility of increased or decreased propagation delays through a device can be represented.

1.3.2. Developing a test

The most common method of developing a test for a logic circuit combines the intelligence of the human-being (the designer or a specialist test programmer) with the computational ability of the computer. The test programmer defines a sequence of inputs to be applied to the circuit and the corresponding outputs. The computer runs a program called a fault simulator which takes descriptions of the circuit and test waveform, together with a list of the faults to be considered and returns an accurate assessment of the test's performance (amongst other information — see Section 1.3.3).

Ways of quantifying test performance and the use of fault simulators are considered in more detail in the following section; here the task of the test programmer in defining the test waveform is considered. First, however, it must be mentioned briefly that computer programs do exist which can automate this task for certain, relatively simple, types of circuit or for circuits designed using highly structured design methods which will guarantee successful automatic test pattern generation. An example of a design methodology that guarantees fully automatic test generation is scan design, which is discussed in detail by Bennetts (1984). Until recently, these structured design methods have only been able to be used for integrated circuit design. The publication of ANSI/IEEE Std 1149.1 has, however, allowed similar techniques to be used at the board level for products constructed from compliant components (see Chapter 3).

How then does the test programmer set about producing a test for a given circuit? Probably there are about as many answers to that question as there are test programmers, so the approach described here — the so-called 'functional' approach — should only be thought of as an example.

The aim of the functional approach is to attempt to get each of the identifiable functional blocks in the circuit — for example, the more complex integrated circuit packages on a printed circuit board — to perform its intended function, and to allow this to be observed at the circuit's normal operating outputs. For example, counters in the circuit will be made to count, shift registers to shift, and so on.

This task is not, unfortunately, as easy as it might appear. If we consider the **device-under-test** (DUT) shown in Figure 1.5, which we will assume to be embedded in the circuit, there are two tasks that must be accomplished.

Firstly, some sequence of values must be applied to the DUT inputs to stimulate it into performing its function. If we consider the DUT to be a counter, then it must be supplied with clock signals and appropriate enables, etc. All these signal values will need to be derived by changing the circuit's 'primary' inputs — the points to which the tester will have access — and propagating the changes through other devices in the region 'A'. Propagation of the changes may be extremely difficult, since the connections between the devices in region 'A' may be such as to cause the required values to be destroyed before they arrive at the DUT.

Secondly, to complete the test of the DUT, its response must be made observable to the tester; that is changes at its outputs must be propagated through devices in region 'B' to the circuit's 'primary' outputs. Again, this will not necessarily be as easy as it might appear, since conflicts

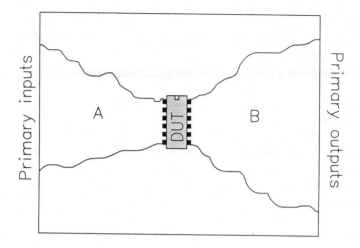

Figure 1.5 The functional test development approach.

may occur on the way. There is the additional problem that the values which have to be set on the devices in region 'B' to allow the DUT's response to be made visible have to be set by applying changes at the circuit's primary inputs, giving rise to the possibility that conflicts may occur between the inputs needed to cause the DUT to operate and those needed to make its operation apparent at the circuit's primary outputs.

To summarize, the two tasks involved are, firstly, control of the DUT inputs and, secondly, observation of the DUT outputs, both of which must be accomplished through the normal circuit connections. The ease of accomplishing these tasks is referred to as the **controllability** and **observability** of the connections and devices in the circuit. These parameters will be mentioned again in the chapters which discuss methods of designing circuits to be more easily testable.

1.3.3. Evaluating test performance

To determine if the performance of the test is sufficient, we need to be able to determine if it detects the target faults for the circuit. This is normally done using a fault simulator which introduces each fault into a simulation of the circuit's behaviour. If the fault causes a change that could be observed by the ATE, then it is deemed detected.

Fault coverage

The measure of test performance is its **fault coverage**. Fault coverage is expressed as the percentage of modelled faults which the test has been shown to detect. Commonly, organizations will set a target of 95% of all stuck-at faults (plus selected open-circuit and bridging faults) below which a test program is unacceptable, although a figure much closer to 100% is desirable. The reasons for accepting a lower figure include:

❏ Certain faults, such as s-a-1 on a connection which is normally tied to logic 1, cannot be detected since they do not change the logical behaviour of the circuit. Such faults may, however, degrade the performance of the circuit — for example, through changes in noise immunity, power dissipation, and so on.

❏ Poor testability may make it very difficult for tests to be produced for some faults within the budget (time and cost) set for the work.

For a figure for the fault coverage of a test to be useful, the set of faults to which the figure relates must be defined. To yield an effective test, the target fault set should include all single stuck-at faults, open-circuit faults, and bridging faults between adjacent pins or tracks.

High fault coverage (as close to 100% as possible) is essential, since faults that the test does not detect may lead to reliability problems when the circuits are installed in working systems, with a consequent loss of the company's reputation for quality (see also Section 1.2.3). Many companies set a minimum acceptable fault coverage of 95%, with the target fault set being all single stuck-at-0 and stuck-at-1 faults.

Fault simulation

As was mentioned briefly before, a fault simulator is a complex computer program used to predict the fault coverage of a test program. It requires three principal computer readable inputs (Figure 1.6):

❏ a description of the circuit diagram (often called a netlist)
❏ a specification of the test waveform
❏ a list of target faults

Once these inputs have been prepared, the simulator can be run and a set of results obtained. Typically, the results will include a list of the faults detected by the test, those not detected, and data to help in locating the source of any fault (for example, a diagnostic dictionary and data for guided probing — see Section 1.5). This process can be extremely expensive and must be carefully planned. For example, fault simulation run times for large, complex circuits are usually measured in central-processor-unit- (cpu-) days, even when using highly-tuned simulation algorithms running on high-performance computers.

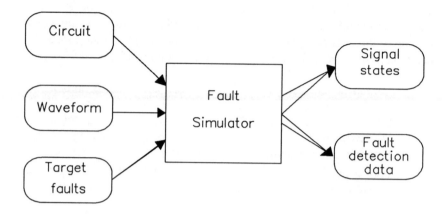

Figure 1.6   The fault simulator.

If the simulation shows that there are faults remaining to be detected, then the test programmer will enhance his test program to attempt to test for them. This cycle — test definition, fault simulation, test definition, etc. — may have to be repeated several times until the performance of the test becomes adequate (Figure 1.7).

1.4. Test application

1.4.1. Automatic test equipment

The automated test systems used for production and maintenance testing come in several varieties.

Integrated circuit testers

Integrated circuit testers can be found both on a semiconductor production line, checking the quality of components as they are manufactured and packaged, or in the incoming goods department of a systems company, where they screen components before they are assembled into hybrids, boards, and so on.

As integrated circuit complexities and operating speeds have increased, general purpose IC testers have become extremely expensive — with costs in excess of $1M being commonplace. Less sophisticated or less generally applicable types of IC tester are available, for example to test prototype IC designs or to test particular types of device (for example, RAMs). These systems can be significantly cheaper, but the performance and range of possible measurements may be limited.

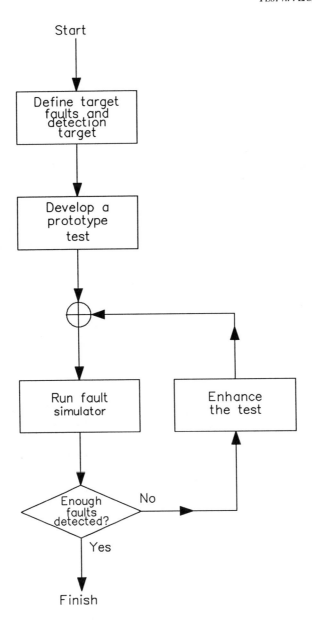

Figure 1.7 The test development process.

Bare-board testers

Bare-board and backplane testers serve an important function in the production of electronic systems. Perhaps the simplest type of tester, they are used to check the integrity of wiring contained on printed circuit boards, or in backplanes or wiring harnesses. The objectives are to confirm that all wanted connections are present, and that no unwanted connections exist.

The tester design is very simple since it has only to allow for resistance measurements to be performed between pairs of contact pins. However, the number of connections can be vast — for example, testers are available that allow simultaneous contact to up to 15,000 points.

In-circuit board testers

The advantages of the bed-of-nails — an array of pins on to which the product is placed during testing that gives access to many internal connections simultaneously — are exploited to the full by the in-circuit board tester. The idea is to provide direct access from the tester to each connection of each component in the circuit and to use these connections to test for the correct operation of each component in isolation, and of the links between them. The assumption is that the majority of the faults that are likely to be introduced during production can be detected in this way. The benefit is that test development costs are low, since tests for a device type can be reused on other products containing the device.

While this type of tester has become increasingly popular in recent years, primarily because the development of test programs for it is relatively inexpensive, it is important to note two points. Firstly, the tests which in-circuit testers apply cannot detect certain faults that can arise through incorrect interaction between components. Secondly, the question of whether the technique may actually damage the components being tested remains unresolved. The problem is that, in order to apply a test to the inputs of one component, the outputs of others will have to be overdriven if they are not in the required logic state. The overdriving process causes a significant amount of current to flow into the output stages of the devices driving the network, which may cause damage if not carefully controlled.

The small example circuit in Figure 1.8 will be used to illustrate the overdriving process. During the test of component G3, components G1 and G2 are powered up and, as a result of signals at their inputs, would normally be driving wires N1 and N2 to 1 and 0 respectively. To test G3 wires N1 and N2 must be set to both 0 and 1 as shown and the ATE does this through the bed-of-nails by supplying sufficient current to force the wires to the required test values, irrespective of the values being driven by G1 and G2. During overdriving the current supplied by the ATE flows mainly through the output stages of G1 and G2 and, if testing continues for sufficient time, this may cause damage. The facility to place a component's outputs in a high impedance state while the components it drives are tested eliminates the possibility of damage to the IC from overdriving.

Another problem area for in-circuit testing arises from the need to make contact to every connection on the board. This may require 1000 or more contacts to be made to the board through a bed-of-nails. For **dual-in-**

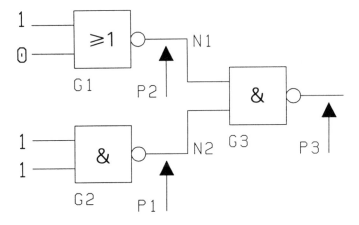

Figure 1.8 In-circuit testing.

line/plated-through-hole (DIL/PTH) technology this degree of access is achievable; however, for miniaturized **surface-mount technology** (SMT) the need for such extensive access can become a significant problem. Cluster/functional testers provide one way of avoiding the worst of these difficulties.

Functional board testers

The functional board ATE works by applying a sequence of logic value changes, the test program, to the product and comparing its observed response with that which would be expected if it were fault-free. The test program is carefully designed for each product with the aim of exercising every part of the circuit. Both the test program and the expected response are held in the memory of the test system.

Normally, both the application of the test program and the observation of the product's response would be done using the product's normal operating connections. However, in some cases the test program may be made more effective by allowing the tester direct access to internal parts of the circuit, either through the use of hand-positioned probes and clips or via a bed-of-nails. Such techniques are used extensively when diagnosing faults in circuit cards, as will be discussed in Section 1.5.

Cluster/functional board testers

Cluster testers provide an effective combination of the in-circuit and functional test approaches. In cluster testing (also known as function testing) groups of components performing an identifiable function within the complete design are tested independently of surrounding circuitry. For example, a microprocessor board will contain functions such as RAM, **read-**

only memory (ROM), communication interfaces, the processor, and so on. Some of these functions are implemented in a single component (for example, the processor), while others require groups of components (for example, the RAM).

The ability to test components in small groups with a clear function allows test development costs to be reduced and, where a function is used in several products, for the test to be reused (as for in-circuit testing). Unlike in-circuit testing, however, defects due to improper interaction between components will be detected. The fact that groups of components are tested, rather than a single component, means that diagnostic tools must be provided (see Section 1.5).

1.4.2. Problems encountered in test application

As with test development, the process of applying a test to a circuit using an ATE can be made more difficult and time-consuming if certain problems are encountered, due both to the electrical and physical design of the circuit.

Considering first the circuit's electrical design, the problems are in the main due to difficulty in controlling the circuit from the ATE. One example is with circuits containing on-board clocks. In such cases it is necessary for the ATE to be able to synchronize to the on-board clock if the circuit is to be tested at all, and (preferably) for the ATE to be able to substitute for the on-board clock during testing. This latter approach allows testing to proceed at the ATE's own speed. Other similar examples of difficulty exist, primarily due to problems in matching the timing of the circuit to that of the ATE. In many cases, the operating speed of the ATE will be considerably lower than that of the circuit being tested, and this should be considered when the circuit is designed. It is almost certain that circuits which depend on critical timings for their successful operation will not be able to be properly tested.

The other area in which problems can arise is in the physical design of the circuit or product — the way that printed circuit boards are designed, and so on. Of particular interest is the ease with which the circuit can be connected to the tester — obviously the more difficult the connection is to make, the slower the process will be.

1.5. Fault diagnosis

After the first application of the test, the ATE will be able to say either that the circuit is fault-free (in which case no further action needs to be taken) or that it contains one or more faults. The next step depends on the type of product and the type of tester being used.

For IC testing, diagnosis is frequently not required, since repair is not practical for many component designs. However, some devices (e.g., RAMs) allow repair by including 'spare' logic blocks within the design that can be switched into the circuit in place of faulty blocks. Diagnosis may also be needed to isolate the cause of the fault to allow modification of the design such that the problem is avoided or to tune the production process.

For board testing, diagnosis is implicit when using an in-circuit test system since the test is applied to one replaceable unit (chip) at a time. For functional and cluster/functional test systems, however, further work must be done to locate the fault once it has been detected. Two methods of automated diagnosis are possible — the use of a fault dictionary, or guided probing — and, while they can be used separately, a combination of them is likely to be more efficient. These two techniques are discussed briefly below.

1.5.1. Fault dictionaries

The fault dictionary is prepared as a by-product of the fault simulation process described in Section 1.3.3. It is a reference table which, for example, gives a list of the faults that will be detected at each step in the test program, organized according to the particular output of the circuit at which the fault becomes apparent. In many cases, the ATE will stop running the test on the first step at which a fault is detected, in which case the fault dictionary will list those faults which are first found at each step.

It is not uncommon for each line of the dictionary to include just one or two faults, so the dictionary can be used to significantly restrict the area of the circuit that needs to be examined for exact diagnosis using the guided probe.

1.5.2. Guided probing

The guided probe is an additional source of information, and can be used to examine each connection in the circuit as if it were a direct connection to the ATE. The fault-free behaviour of all internal connections will have been evaluated during fault simulation and filed in the ATE's memory, for later comparison with the performance of a circuit under test.

During guided probing, the ATE will direct the operator to place the probe at specific points in the circuit in an ordered manner, as illustrated in Figure 1.9. In the absence of a fault dictionary, probing will start from the circuit output at which the fault was first detected. The probe is then placed in turn on the input connections to the device feeding that output and the test program re-run. The output from each connection is compared with that stored on the ATE and, as a result, 'bad' or 'good' flags can be associated with each of the inputs to the device. The process is repeated, tracing the 'bad' connections back through the circuit, until either:

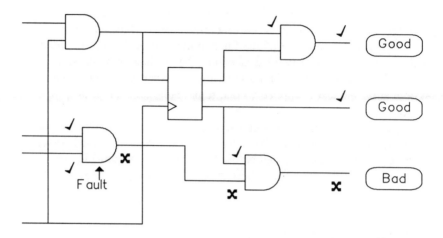

Figure 1.9 Guided probing.

(1)   A device is found which has a 'bad' output but 'good' inputs. This may be caused by a faulty device, by a fault on the output connection (for example, a short circuit or a stuck-at fault — but not by an open-circuit fault), or by a fault at the input of a device driven from a 'bad' signal.

(2)   A connection is found with a 'good' signal at the driving end, but a 'bad' signal at the receiving end. This could be caused by an open-circuit fault, for example.

The basic procedure described above will almost certainly be enhanced to improve the accuracy of the diagnosis. For example, both the driving and receiving ends of a suspect connection may be probed so that open-circuit connections can be diagnosed. Also, if a fault dictionary is available, then probing may start at some point within the circuit rather than at the circuit outputs, thus cutting out much unnecessary activity.

A further refinement is the use of a current-sensitive probe to resolve the locations of certain types of fault on bus-structured circuits. The probe allows the location of the fault to be determined, even though its effect is evident on all parts of the bus.

1.5.3. Problems encountered during fault diagnosis and repair

Problems encountered during diagnosis are primarily due to the physical design of the product being tested — for example, problems in placing probes on connections in the circuit because devices are too closely spaced. Other problems can be caused by the layout of the circuit which can impede diagnosis by making it difficult for the operator to locate or access the points he is asked to probe.

1.6. Design-for-test

The goal of design-for-test is to ensure that a completed product design can be economically tested, both following manufacture and during its operational life. The following sections provide a brief introduction to the aims and scope of design-for-test.

1.6.1. Do you want to create a 'quality' design?

Would you like to feel that you have created an excellent product and not have the manufacturing and repair organizations complain about poor testability?

If your answer is yes, read on. This book has the information you need to create a testable product that is matched to test capability in a wide range of manufacturing and repair operations.

1.6.2. Design-for-test starts when design work starts

Designs which work as bench-top prototypes may not necessarily be capable of volume manufacture or be economically supportable, since manufacturability or maintainability may not have been properly considered. Manufacturability issues include the suitability of the product for auto-insertion of its components, the suitability for wave soldering, the provision of adequate timing margins, and the ease of testing the assembled product. Maintainability issues include reliability, the ease of testing in the field and during repair, and the ease of diagnosing faults. Indeed the general experience throughout the electronics industry is that the designer must specifically target manufacturability, maintainability and other issues from the outset if a product is to be a success, both technically and commercially.

Testability — the ease of testing a product and, when required, locating faults — is a key contributor to both manufacturability and maintainability. Many electronics companies now state that test costs account for of the order of 50% of the total life cycle cost, so testability can have a significant impact on a product's commercial viability.

The best way to achieve adequate testability is to include testability as a design objective from the outset. Designing testability into a product after the initial design is complete is difficult, a waste of valuable resources, and will, in general, produce a less than satisfactory result.

1.6.3. What aspects of a design impact its testability?

From the preceding sections, it can be seen that testability impacts two facets of product design:

❐   *Circuit design.* Here design-for-testability will permit easier test development and may give shorter test application times.

❐   *Physical design.* For example, design-for-testability will ensure that sufficient access is available to allow internal connections of the product to be examined during fault diagnosis.

1.6.4. Testability isn't free

To make a product testable, circuitry or other features must usually be added to those needed to realize the intended function. Therefore the benefits of having a testable design must clearly be viewed against the costs of achieving testability. Costs (like benefits) fall in two areas:

❐   *Non-recurring costs.* Non-recurring costs arise through increased design time, both to include testability features in the design and in the testability assurance process. Generally these costs will be a small fraction of the complete design cost, particularly if the designer is properly trained in design-for-testability. Non-recurring costs are offset by reductions in the start-up costs of testing (e.g., in test development) and savings in recurring test costs (e.g., the time to test and diagnose the product).

❐   *Recurring (per item) costs.* These arise during manufacture and/or maintenance. Recurring costs result from increased board or chip size, added logic or components needed to achieve testability, and so on. These costs will be offset by benefits such as reduced test and diagnosis time on the ATE (giving increased throughput).

The objective is to give a reduced overall life cycle cost for the product, not to minimize the localized costs of one aspect of the product's life. Testability may not be free, but any investment must produce a return.

The benefits of having a testable design must clearly be viewed against the costs of achieving testability. Costs (like benefits) fall in two areas: non-recurring costs during product development and recurring (per item) costs during manufacture and/or maintenance. Non-recurring costs arise through increased design time, both to include testability features in the design and in the testability assurance process. Generally these costs will be a small fraction of the complete design cost, particularly if the designer is properly trained in design-for-testability. Recurring costs result from increased board or chip size, added logic needed to achieve testability, and so on. Clearly an objective during design-for-testability is to keep the recurring costs to a minimum — but it must be recognized that testability may not be free.

References

Useful tutorial texts

The following is a short list of tutorial texts which contain a more complete discussion of electronic test technology (including test generation and design-for-testability) than that included in this chapter.

Agrawal V.D. and Seth S.C. (1988). *Test Generation for VLSI Chips*. Los Alamitos, Calif.: IEEE Computer Society Press.
Bennetts R.G. (1981). *Introduction to Digital Board Testing*. New-York & London: Crane-Russak/Edward Arnold.
Bennetts R.G. (1984). *Design of Testable Logic Circuits*. London: Addison-Wesley.
Miczo A. (1986). *Digital Logic Testing and Simulation*. New York: Harper & Row.
Parker K.P. (1987). *Integrating Design and Test — Using CAE Tools for ATE Programming*. Los Alamitos, Calif.: IEEE Computer Society Press.

Papers

Eldred R.D. (1959). Test routines based on symbolic logic statements. *Journal of ACM*, **6**(1), 33-36.
Reinerstein D.G. (1983). Whodunit? The search for the new-product killers. *Electronic Business*, (July), 62, 64, 66.
Williams T.W. and Brown N.C. (1981). Defect level as a function of fault coverage. *IEEE Transactions on Computers*, **C-30**(12), 987-988.

CHAPTER 2.
Design-for-Test Techniques

2.1. Introduction

Chapter 1 highlighted several problems that can arise when testing an integrated circuit or loaded board. Key causes of these problems are:

(1) *Complexity*. The difficulty in generating a test is related both to the size (for example, number of gate-equivalents) and complexity (for example, amount of feedback around or cross-connection between logic blocks). For combinational circuits, test generation costs vary on the order of N^2, where N is the number of gate-equivalents in the design (Goel, 1980). For sequential circuits, test costs are further increased by the presence of stored-state devices and feedback.

(2) *Speed*. For state-of-the-art ICs or boards, the maximum operating speed is likely to exceed that of the automatic test equipment (ATE)

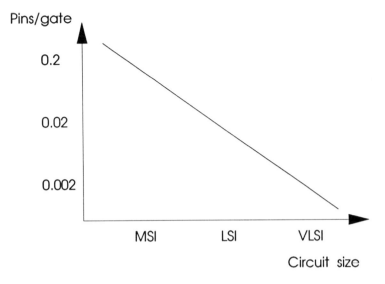

Figure 2.1 Variation of pin count with IC complexity.

used to apply tests. The ATE is built using yesterday's technology, but is expected to test today's products.

(3) *Access.* During functional testing, most (if not all) connections between the ATE and the IC or board are made through the normal inputs and outputs (the package pins or board connectors). While the complexity of ICs and boards is rising rapidly, the number of external connections is relatively static (e.g., as shown in Figure 2.1). Therefore, more and more test data must be transferred through a limited number of connections from the ATE to the unit under test (UUT). This causes a bottle-neck, increasing test time and, in consequence, reducing ATE throughput.

(4) *Miniaturization.* Through use of surface-mount assembly techniques, the geometries of loaded boards can be reduced considerably. The result, however, is that access to internal chip-to-chip interconnections on the board using bed-of-nails or guided probes becomes difficult. Unfortunately, probing is an inherent feature of in-circuit and functional testing.

The use of design-for-test techniques during the development of a new circuit design can reduce these problems and, as a result, reduce the cost of testing the circuit in production or during field service.

In this chapter, an overview will be given of the principal design-for-test techniques for digital circuits — both for ICs and for loaded boards. The intention here is to give readers an understanding of what each technique involves. The techniques that are of most value in the design of loaded boards — design-for-test guidelines and boundary-scan — will be discussed in more detail later in the book. The remainder are most widely used in the design of ICs. In these cases, the aim of this chapter is to give readers sufficient information to allow them to converse with the designers and vendors of ICs — not to enable them to design the ICs themselves.

For convenience, the design-for-test techniques are introduced in an historical sequence.

2.2. Do nothing

Of course, the designer could decide to ignore testability completely — to concentrate solely on meeting the functional requirement for the design.

> 'Nobody worried about design-for-test when designing boards with small and medium scale ICs, so why bother now? The test engineering department could generate tests for everything we sent them. Can't they still do that for designs that use VLSI ICs?'

The problem is that test costs rise exponentially as the complexity of a circuit rises. Double the circuit complexity and, without design-for-test, test costs may quadruple — that is, if a test can be generated at all.

True, in the 1970s, when ICs and loaded boards contained only simple circuits, test engineers were able to create thorough test programmes for every design they received. They did not have to become involved in the design process themselves because, although it would be more difficult to create tests for some designs than for others, the cost of the work was small compared to the design cost — even for those designs that were quite difficult to test.

Organizations and the people in them adapted to this scenario:

❒ Highly-separated design and test teams were created.

❒ The 'over the wall' mentality evolved. Designers created the product and refined it up to a point where it could be passed on to production engineering. Production engineering dealt with the task of turning the prototype design into something that could be manufactured and tested. Test engineering could be staffed with relatively low-grade labour. Highly-trained engineers were needed for design; failed designers went into test engineering! (Needless to say, the author does not share this view!)

Some companies are still suffering from these attitudes and organizational structures today.

For very-large-scale integration (VLSI) ICs, and boards that contain them, test engineering is a function that requires highly-skilled staff who must be closely involved in the design process from the outset. Today, a company that omits consideration of testability from the design process is taking an enormous commercial risk:

❑ The usual result of poor testability is that the test engineering team will not be able to create a sufficiently thorough test programme within the time and budget allowed. This is illustrated in Figure 2.2.

As testability reduces, the fault coverage achievable on a fixed budget reduces. This is important, because time-to-market pressures normally prevent more time and budget being allocated to overcome test problems (Reinerstein, 1983).

❑ With an inadequate test programme, a greater number of defective parts will be shipped to customers — the tests needed to highlight the presence of their faults just aren't in the test programme. Figure 1.2 in Chapter 1 showed the relationship between shipped product quality, manufacturing yield, and test programme effectiveness.

❑ Customers find the defects missed by the test — Murphy's Law applies (if anything can go wrong, it will — usually at the most inconvenient time).

❑ Companies with a reputation for poor quality lose business!

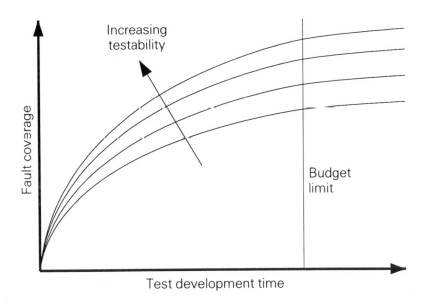

Figure 2.2 The effect of testability on test development time.

2.3. Design-for-test guidelines

Design-for-test guidelines are lists of *dos* and *don'ts* generated by test engineering departments. They record ways of designing circuits shown by experience to result in reduced test costs (the *dos*) and ways of designing circuits that result in increased test costs (the *don'ts*). In effect they provide a feedback path from test engineers to designers that, over a period of time, will result in better (that is, more testable) designs emerging from the design department.

Part 2 of this book contains an extensive list of design-for-test guidelines for use during board design. Many of these guidelines are also useful in the design of ICs. To illustrate how design-for-test guidelines might be applied, consider Figure 2.3.

Perhaps the single most important guideline is 'Ensure that the circuit can be initialized — quickly and easily.' The reason for this guideline is that no testing can be done until the circuit has been placed in a known starting state. After this time, inputs can be applied to the circuit and its response can be observed. Prior to initialization, observation of the outputs of the circuit is of little value, because the circuit is simply moving from one indeterminate state to another.

The example circuit of Figure 2.3 is difficult to initialize. Inspection of the design shows that the clear input to IC13 is supplied from an output of IC6. IC6 has its preset input tied high (inactive), while its clear is supplied from IC13, via an inverter. Therefore, to reset IC13 (an 8-bit shift register) it is first necessary to clock it until a logic 1 is propagated from its input (INa) to output Q8. As soon as Q8 is set to 1, IC6 is cleared and IC13 itself is reset. Note that the clear input to IC25 is also supplied from IC6. Therefore, clocking of IC13 is synchronized to the arrival of a logic 1 at output Q8 of IC13.

The circuit is, in fact, an asynchronous data receiver, similar to the receive portion of a universal asynchronous receiver/transmitter. Data packets are received at the serial command input and are clocked into the shift register. The clock for the shift register is synchronized as closely as possible to the arrival of the start bit (a one).

Initialization can be achieved by entering a known 8-bit data stream and watching to see when it appears as a set of parallel bits at the circuit's parallel outputs. While a test system can be programmed to do this, the problem is that none of the most widely used simulators can simulate the behaviour of this circuit during initialization. The majority of simulators use a single value (X) to indicate that the state of a signal is unknown. Therefore the simulated circuit will simply move from one unknown state to another as data is clocked in — initialization will never be achieved. The point here is that initialization is difficult because:

(1)   a great deal of thought is required to analyse the circuit and work out how it can be set to a known initial state; and

(2)   in the majority of cases, the initialization sequence cannot be simulated — preventing accurate fault simulation during this vital stage of testing, for example.

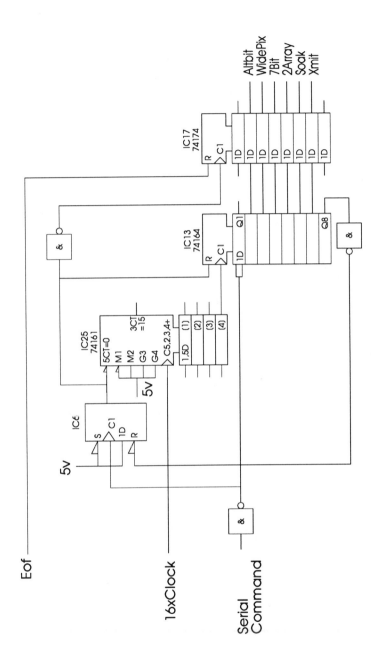

Figure 2.3 An example circuit.

Initialization can be achieved easily by adding an external reset input to the stored-state devices, gated in with the clear signals required for normal operation.

Of the design-for-test techniques that will be introduced in this chapter, the use of guidelines allows the widest range of test problems to be dealt with. While scan design and self-test concentrate on the logical design of the circuit, design-for-test guidelines can be used to tackle problems that arise from the physical design of the board — for example, difficulty in probing because probe targets are too closely spaced.

2.4. Scan design

The scan design technique is widely used in the design of **application-specific ICs** (ASICs) and very-large-scale-integration ICs (VLSI ICs).

2.4.1. Test generation problems

In a combinational logic circuit, the states of the outputs are determined solely by the signals applied at circuit inputs. As a result, test generation is (relatively) straight-forward — certainly, it is sufficiently easy to allow computer programs to be written to perform the test generation task.

In contrast, test generation for stored-state logic circuits of the complexity typically found in industry is an extremely complex task that cannot be automated unless a significant investment is made in design-for-testability. (The use of design-for-test guidelines alone is unlikely to be sufficient to permit automated test development.) Why is this so? There are two principal reasons: stored-state devices and feedback.

Stored-state devices

For a stored-state circuit, the output is determined not only by the signals being applied at the input at that time but also by previous input signals, a processed form of which is held in the various stored-state devices. To set any signal to a value required during testing, it is therefore necessary to compute a *sequence* of input stimuli. A further sequence of stimuli will be needed to make the state of any given part of the circuit visible at the circuit outputs.

Feedback

Feedback comes in two types. There are *local* feedback paths built into each flip-flop or latch. *Global* feedback paths are those that are external to the flip-flops or latches themselves (see Figure 2.4).

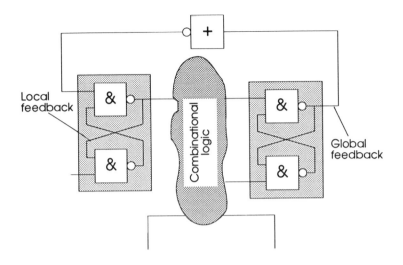

Figure 2.4   Local and global feedback.

In the absence of global feedback, test generation for a stored-state circuit is only slightly more complicated than that for combinational circuits. In effect, the flip-flops or latches result in pipelining of input signals, delaying the time at which they arrive at any given point by an appropriate number of clocks. (In the purely combinational circuit, the delay would be limited to that caused by the signal propagation characteristics of the various gates.) The signal at any node is determined by the values at *other* nodes at previous times — but not by the value previously held at the node itself.

The presence of global feedback severely complicates test generation. Now, the signal present at any node may be dependent on a previous state of the same node. The test generator (human or computer) must not only compute the state to which a node must be set, but also the time that the node must be in this state. The node may be set to different states at different times, but not to different states at the same time, so a record must be kept of state assignments against time. The result is a significant increase in the amount of effort required to generate tests.

2.4.2. The principle of scan design

For loaded boards, a frequently-used method of solving test problems caused by complex VLSI ICs is to unplug these components from the board during functional testing. This allows the test system to control and/or observe the signals that would otherwise flow into or out of the complex chip. (Note that physical removal of the chip is not necessary in situations where all the chip's output pins can be placed in a high-impedance state.) These signals become primary inputs and outputs for the circuit for the duration of the test and are referred to as pseudo primary inputs and pseudo primary outputs.

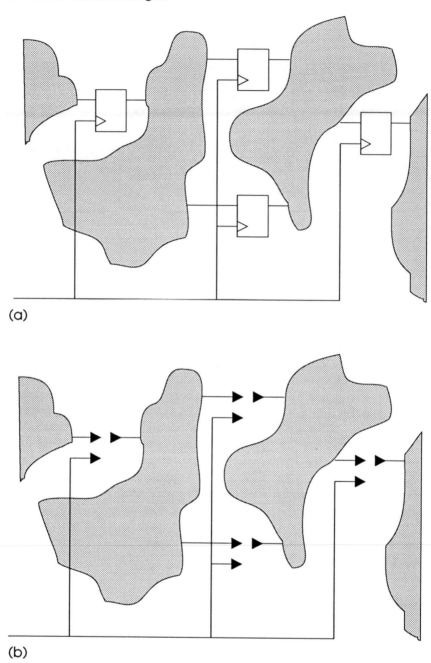

(a)

(b)

Figure 2.5   Removal of flip-flops from a stored-state circuit.

Imagine applying this process to a simpler circuit consisting of combinational logic and flip-flops, with the flip-flops being the 'complex' devices. Removal of the flip-flops would leave a purely combinational logic network, as shown in Figure 2.5. Tests for this network could be generated fully automatically at low cost. The flip-flops could be tested independently of the combinational logic before being returned to the circuit.

Scan design techniques provide a logical (rather than physical) means of removing the flip-flops and latches from a stored-state circuit design, with the benefit that test generation can be fully automated (Eicherberger and Williams, 1977).

2.4.3. Shift register scan

There are several forms of scan design (McCluskey, 1984). The most common form requires that all the flip-flops and latches in a circuit design are connected to form one or more shift register paths when a special test mode is selected. In Figure 2.6b, this is achieved by provision of a multiplexor at the data input to each flip-flop. One data input to each multiplexor receives the signal previously fed to the flip-flop's data input (Figure 2.6a), while the other is fed by the output of the preceding stage in the shift register chain (or, in the case of the first flip-flop, from a serial input, Scan-In). The control inputs to the multiplexors are fed from a dedicated test control input, Test-Mode, and the data output from the last flip-flop in the chain is fed to a serial output, Scan-Out.

Testing of the modified circuit proceeds in two stages: shift register test and combinational logic test.

Shift register test

Test-Mode is set to select shift register operation of the modified flip-flops. Data is clocked in through Scan-In and appear at Scan-Out after an appropriate number of clock pulses has been applied. A data sequence of the form:

$$01001100011100001111...$$

tests that:

(1)   each flip-flop can be set to both 0 and 1;

(2)   each transition in state is possible (0 to 1 and 1 to 0); and

(3)   there are no 'pattern-sensitive' faults — for example, faults that would prevent a change of state following a prolonged sequence of 1s or 0s.

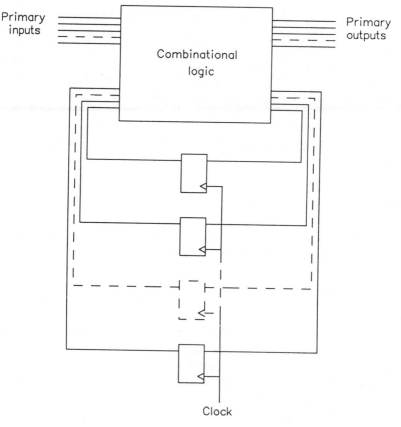

Figure 2.6   Shift-register-based scan design — (a) Huffman model.

Following this test, the only faults in the storage devices that have not been tested are those on the connections to and from the combinational logic.

Combinational logic test

This stage tests for the remaining faults in the flip-flops and for faults in the combinational logic.

Tests for the combinational logic network are generated using an **automatic test pattern generation** (ATPG) package. For each of the test vectors generated, the following procedure is used:

(1)   Data is applied to those inputs of the circuit that are directly accessible.

(2)   The remainder of the input test pattern is shifted into the various flip-flops by selecting test mode (Test-Mode = 1) and applying clock pulses.

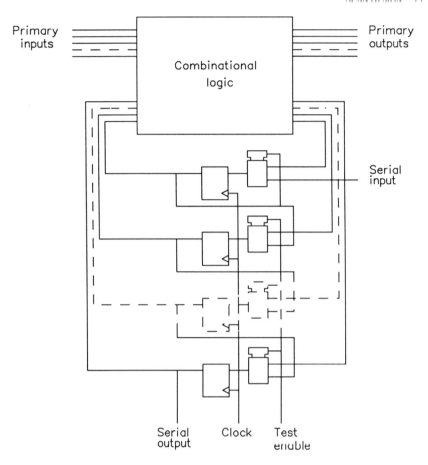

Figure 2.6 (cont.) Shift-register-based scan design — (b) Scan design.

(3)   When shifting is complete, the circuit is placed in its normal mode (Test-Mode = 0). Those outputs from the combinational logic that are directly observable are checked against the expected values.

(4)   One clock pulse is applied. This causes the remaining outputs of the combinational logic to be loaded into the flip-flops to which they are fed.

(5)   The circuit is set to test mode (Test-Mode = 1) and clock pulses are applied to shift the captured data out of the circuit through Scan-Out. Following each clock, the signal at Scan-Out is compared to that expected at the appropriate combinational logic output.

Steps (2) and (5) may be merged, with the results of one test being shifted out as the stimuli for the next are shifted in.

There are several alternative forms of scan design based on the use of shift registers. Of these, the **level-sensitive scan design** (LSSD) technique is the most widely used. LSSD is based on the use of latch-based shift register stages comprising a pair of latches (a master and a slave) controlled by independent non-overlapping clocks (Figure 2.7). A significant advantage of LSSD is that it eliminates a number of timing problems that can arise in use of the multiplexor/flip-flop design presented earlier.

Further details of LSSD and a set of design rules for a scan testable circuit can be found in Bennetts (1981).

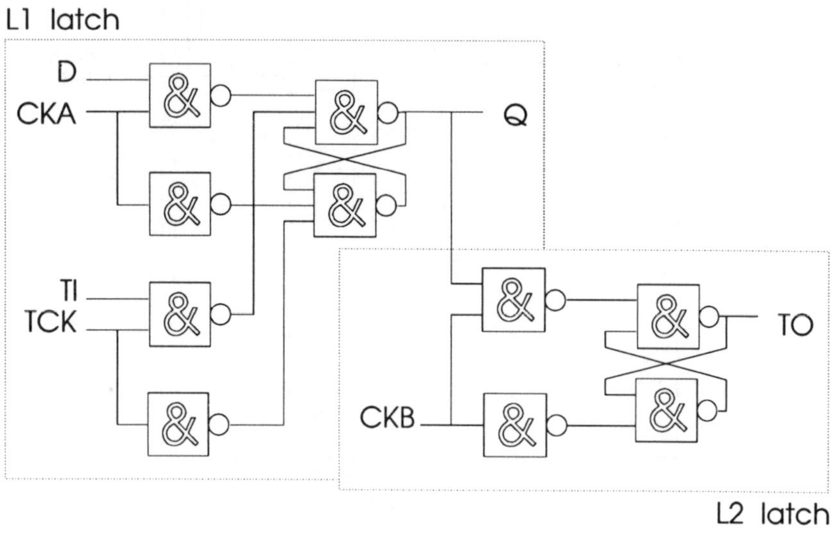

L1 latch

L2 latch

Figure 2.7 An LSSD shift register latch.

2.4.4. Random-access scan

In random-access scan, individual latches or flip-flops are built into a RAM-like structure that allows their state to be individually written (controlled) and read (observed) (Ando, 1980). The resulting design is shown in Figure 2.8.

In Figure 2.8, the circuit is again divided into two parts — the combinational logic and the stored-state devices. The stored-state devices are level-operated latches implemented as shown in Figure 2.9. The scan data input for the chip is broadcast to all latches (to SDI). The scan data output of each latch (SDO) feeds onto a wire-AND bus and thence to the chip's scan data output.

For normal circuit operation, the scan clock (SCLK) is held low. Data at the system data input (D) is loaded into the latch when the system clock (CLKn) is high. At least two system clock signals are required (CLK1 and CLK2) and all system clocks must be non-overlapping (that is, only one of them is high at any given time). Latches controlled by one system clock

(CLKn) can only feed latches that are controlled by different clocks (CLKm, $m != n$).

In test operation, the system clock signals are held low. A target latch is selected by feeding an address into the circuit. The address is decoded to set the select input (SEL) of the target latch high, while the SEL inputs for all other latches are set low. This enables the target latch to drive onto the scan data out bus. New test data is written to the target latch from the chip's scan data input by setting SCLK high while SEL is high.

If all latches need to be observed and controlled during application of a test, then each must be addressed in turn. SCLK must be controlled to ensure that data is read from the addressed latch before new data is written.

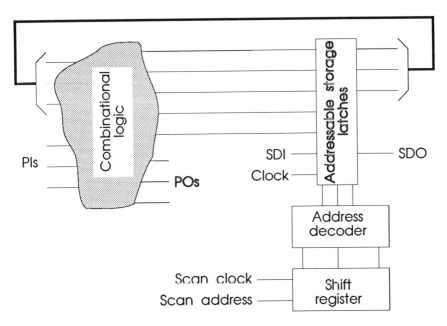

Figure 2.8 Random access scan.

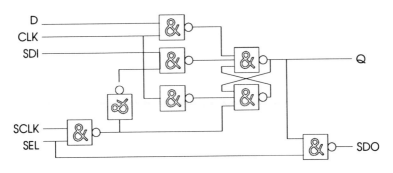

Figure 2.9 Random-access scan latch.

Once data is written into every latch, SCLK is held low while the system clock signals are pulsed in the appropriate sequence. The data in the latches can then be read by holding the system clocks low and sequencing through the addresses.

The advantage of random-access scan compared to a shift-register-based implementation is that it is possible to read and/or write data into individual latches without altering the state of the rest. This can be useful during functional testing or debug.

2.4.5. Partial scan

The scan design techniques discussed so far involve the provision of test access to every latch and flip-flop in a chip design. As a result, a circuit is produced for which automated test generation is guaranteed — regardless of the detail of the design.

The partial scan technique involves the selection of a number of flip-flops or latches to be included in a shift register path through the design. Alternatively, shift-register-based testability improvement cells of the form shown in Figure 2.10 can be introduced into the combinational circuitry to provide access during testing.

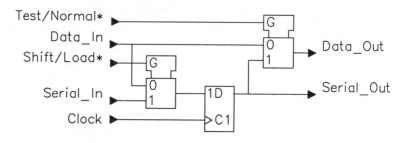

Figure 2.10   Testability improvement cell.

The idea is to provide shift register access to locations in the circuit that are particularly difficult to access, and thereby bring the resulting design within the reach of an ATPG tool. For example, tools are available that identify the nodes, flip-flops, or latches to be included in the shift register path during the ATPG run. Other tools have been proposed that analyse a design before test generation is attempted, producing a recommended set of storage devices for inclusion on the shift register path.

Compared to full scan design, partial scan can be implemented at lower circuit cost. However, more computational effort is required during the design process because the sites for shift register access must be identified. A further limitation is that, while a full scan circuit can readily be 'upgraded' into a self-testing circuit (see Section 2.5), this is not the case for a partial scan design. Problems may also result when a completed design is used as a building block in a later, larger circuit because the locations that need to be scannable to render the original design testable may not be those to which access is required to permit testing of the new circuit.

2.4.6. Boundary-scan

Boundary-scan is a technique that can be used both in the design of large, complex ICs, multi-chip modules, and loaded boards. In 1990, the IEEE approved a standard specification for boundary-scan facilities to be built into ICs (IEEE, 1990) and, as a result, the technique looks set to become one of the major ways of testing loaded boards (including multi-chip modules).

Chapter 3 provides a detailed description of the features specified by the IEEE standard and how they can be used. Therefore, this section is intended only to give a very brief overview of the boundary-scan technique.

To implement boundary-scan, a scan shift register stage is placed adjacent to every input and output of an IC building block (macro) or chip —that is, at the boundaries of the circuit. (Henceforth, the case of an IC will be described. Remember, however, that the technique is equally useful *within* a chip design.) To achieve this, specialized test circuitry may need to be added to the chip between each pin and the logic to which it is connected, as shown in Figure 2.11. These test circuits, called boundary-scan cells, are connected into a shift register path around the periphery of the IC. This is called the boundary-scan path.

An example design for a boundary-scan cell is shown in Figure 2.12. (As will be discussed in Chapter 3, this is typical of the cell designs permitted by the IEEE standard.) Data can flow directly through the boundary-scan cell (from Data-In to Data-Out) when normal operation of the

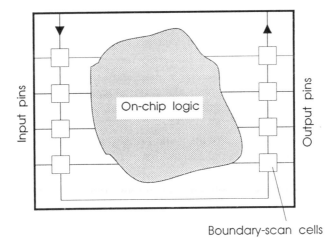

Boundary-scan cells

Figure 2.11 Inclusion of boundary-scan in an IC.

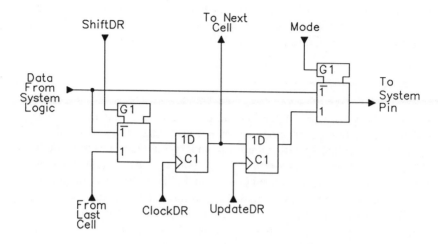

Figure 2.12 A boundary-scan cell.

component is required. During testing, the cells at output pins can be used to drive signal values onto the external network (e.g., the board interconnect), while those at input pins can capture the signals received.

With cells of the design shown in Figure 2.12, testing is performed with the Test/Normal* signal set to 1. Two principal types of test are possible: interconnecton test and chip test.

Interconnection test

Test patterns are shifted into the boundary-scan cells at chip output pins and driven onto the external connection. The results of the test arrive at the input pins of an adjacent chip and are loaded into their boundary-scan cells (Shift/Load* = 0). They are then shifted out for examination (Shift/Load* = 1). By careful selection of test patterns, the interconnections between boundary-scan-testable ICs can be tested for stuck-at, short circuit, open circuit, and other fault types. Figure 2.13 shows a circuit that contains a short-to-ground (stuck-at-0) fault and a wire-OR short circuit fault in the board interconnect (for example, a solder bridge). Table 2.1 shows some test vectors for these faults.

Note that the rightmost bit of the data in Table 2.1 is shifted into the serial input, or out of the serial output, first.

Table 2.1 Example tests for interconnect faults.

Input	Output	
	Expected	Actual
x1x1x0xxxxxx	xxxxxxxx01x1	xxxxxxxx11x0
x0x0x1xxxxxx	xxxxxxxx10x0	xxxxxxxx11x0

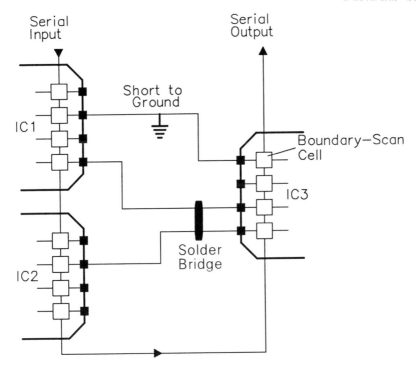

Figure 2.13 Testing for interconnect faults.

Chip test

Figure 2.14 shows a simple IC that contains a NAND gate. To test to this gate, stimuli are shifted into the boundary-scan cells located at the input pins. The result of the test is loaded into the cells at the chip's output pins (Shift/Load* = 0) and then shifted out for examination (Shift/Load* = 1). A set of test vectors for the NAND gate is shown in Table 2.2. As for Table 2.1, the rightmost bit of each data pattern shown in Table 2.2 is shifted into the serial input, or out of the serial output, first.

   If the target chip is scan testable, then operation of its internal scan path can be synchronized to that of the surrounding boundary-scan path during application of the chip test.

   A significant advantage of the boundary-scan technique is that it separates the tasks of chip testing and loaded-board testing. In particular, only a limited knowledge of the chip's function and design is required to allow a high-quality test to be generated for board-level interconnections — indeed, this is possible from a knowledge of the design of the boundary-scan path alone. This is a notable contrast to the case without boundary-scan, where the test engineer needs to know a great deal about the chip's operation to generate either an in-circuit test or a functional board test.

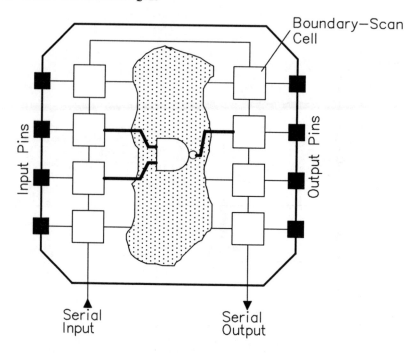

Figure 2.14 Testing on-chip logic.

Table 2.2 Example tests for the NAND gate.

Input	Expected output
x10xxxxx	xxxxx1xx
x01xxxxx	xxxxx1xx
x11xxxxx	xxxxx0xx

2.5. Self-test

For VLSI ICs, the problems of complexity, speed, and access outlined at the start of this chapter become particularly severe. **Built-in self-test** (BIST) techniques offer a good solution to these problems. As will be discussed in Chapter 3, BIST features in ICs can also be accessed at board and system levels provided that a suitable control and access mechanism exists, reducing the cost of testing these higher level assemblies as well as the cost of testing the chip itself.

In BIST, some or all of the function of the ATE is built into the chip itself (Figure 2.15). Additional logic is provided to generate test stimuli and observe test responses — functions that the tester would otherwise perform. To limit the cost of this added circuitry, BIST designs are most often based on the use of pseudo-random testing, in which large numbers of near-random test stimuli are applied to the circuit. The responses of the circuit to these tests are compressed into a signature, consisting of a relatively small number of data bits, that can be fed out of the circuit at the end of the test and checked against the expected fault-free value.

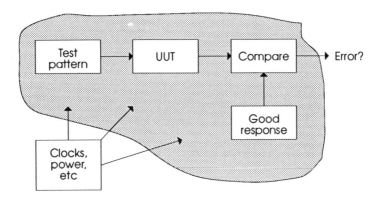

Figure 2.15   Built-in self-test.

2.5.1. Generating the test

Figure 2.16 shows a small **linear-feedback shift register** (LFSR) design that can be used to generate a sequence of pseudo-random test stimuli. Outputs from a number of shift register stages are fed to an exclusive-OR gate network (parity tree) the output of which is fed back to the input to the first stage.

If circuit is initialized to a starting state other than that where every stage is set to 0, then the example LFSR will generate a sequence of patterns that repeats after 15 (2^4 - 1) clocks have been applied. (If the circuit starts in the state where every stage is set to 0, then it will remain in this state no matter how many clocks are applied.) The sequence generated from a starting state where every stage holds a 1 is shown in Table 2.3. Note that the sequence of 1s and 0s generated at any particular output appears to be random. Because the sequence is actually determinate, it is termed pseudo-random. The randomness of the outputs increases as the number of shift register stages in the LFSR increases.

In the general case, an LFSR built from a shift register containing N stages can generate a sequence of length 2^N - 1 if the points from which feedback is derived are chosen carefully. (Tables are available to assist in choosing the correct feedback taps for a shift register of a given length — for example, see Bardell (1987)). Use of a different set of feedback 'taps' may result in a sequence being generated that repeats more frequently than every 2^N - 1 clocks.

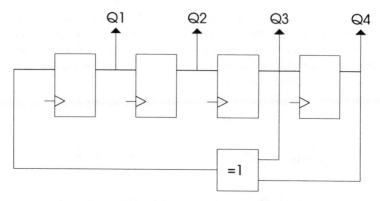

Figure 2.16   Linear-feedback shift register.

Table 2.3   Sequence generated by the example LFSR.

Q1	Q2	Q3	Q4
1	1	1	1
0	1	1	1
0	0	1	1
0	0	0	1
1	0	0	0
0	1	0	0
0	0	1	0
1	0	0	1
1	1	0	0
0	1	1	0
1	0	1	1
0	1	0	1
1	0	1	0
1	1	0	0
1	1	1	0
1	1	1	1

2.5.2. Signature analysis

A signature analyser — that is, a circuit that can compress the sequence of results from a test into a short signature — can be produced by adding an external input to an LFSR as shown in Figure 2.17. The signature generated can be read at the end of the test and compared to that expected from a fault-free circuit. Table 2.4 shows how an example fault-free output from a circuit is compressed into a signature (1110).

Signature analysis was originally developed as a tool to help in the location of faults in digital equipment — for example, as a tool for use during on-site repair (Frohwerk, 1977). Signature analysers can often be found built into ATE systems because they reduce the amount of test-result data that has to be stored. Often, these signature analysers are based on 16-bit shift registers. For BIST applications, the size of the signature analyser would vary according to the length of the test that is applied.

For test sequence of reasonable length (say, more than a few tens of tests), the chance that a faulty input sequence from the circuit under test will produce the expected fault-free signature (that is, the chance that the presence of the fault will be masked) is approximately

$$\frac{1}{2^N - 1}$$

where N is the number of shift register stages in the signature analyser. That is, for a 16-bit design, a maximum of only 0.002% of faults will be masked by the signature analyser. From a practical point of view, it can be considered that masking will rarely occur for signature analysers of length 16 and over.

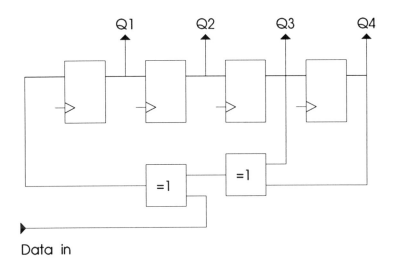

Figure 2.17 A signature analyser.

Table 2.4 Signature generation.

Q4	1	1	1	1	1	0	1	0	0	0	0	0	0	1	0	1	0	1
Q3	1	1	1	1	0	1	0	0	0	0	0	0	1	0	1	0	1	1
Q2	1	1	1	0	1	0	0	0	0	0	0	1	0	1	0	1	1	1
Q1	1	1	0	1	0	0	0	0	0	0	1	0	1	0	1	1	1	0
Data	1	0	1	0	1	1	1	0	0	1	0	1	1	0	0	0	1	

2.5.3. The multiple-input signature register

The signature analyser so far described can be used to monitor a single output stream from the circuit under test. However, circuits typically have more than one output. Either sufficient signature analysers must be provided such that each circuit output stream can be compressed into a signature, or the test must be repeated many times with a signature being taken from one output on each occasion.

The **multiple-input signature register** (MISR) overcomes this problem. As shown in Figure 2.18, an input data stream can be fed in at each stage along the shift register. Therefore, a MISR can monitor as many circuit inputs as the number of shift register stages it contains.

Unless care is taken in the way that circuit outputs are connected to MISR inputs, there is a slightly higher chance of fault masking where a MISR is used. Figure 2.19 illustrates the problem.

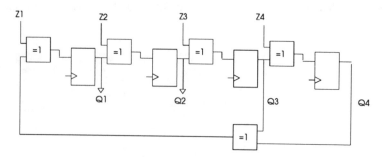

Figure 2.18   The multiple-input signature register.

(a) Waveforms at MISR inputs (shading shows errors)

(b) States of MISR latches (X indicates error)

Figure 2.19   Fault masking in a MISR.

For example if a faulty data bit is injected into a shift register structure built into the circuit under test, then this bit could be presented at two inputs to the MISR — but at different times. When the faulty data bit arrives at the first MISR input, it corrupts the data held in shift register stage 1. Either a 1 is changed to a 0, or a 0 is changed to a 1.

On the next clock cycle, the corrupted data in the MISR will move into stage 2. On the third clock, the corrupted data in stage 2 will be combined with a faulty data bit received at the second MISR input, with the result that a *correct* value is written into stage 3 (a case of *two wrongs make a right*). The fault has been masked.

The problem could be avoided by shifting the input connections one stage towards the right (that is, the input to Z1 is now fed to Z2, and so on, with the input to Z4 becoming applied to Z1). Now, the fault first results in the corruption of the data held in stage 2. The corrupted data is moved into stage 3 on the next clock. On the next clock the corrupted data is combined with the faulty data at the second MISR input and, as before, results in correct data being written into the next stage — stage 4. However, the corrupted data in stage 3 is also fed back to stage 1 on this clock.

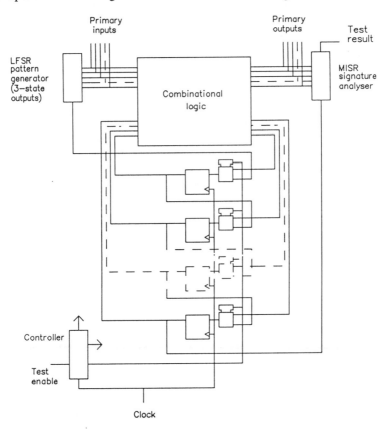

Figure 2.20 A self-testing IC design.

Therefore, while the data in stage 4 does not show the presence of the fault, the data in stage 1 does.

In general, if masking is to be minimized, related outputs from a circuit under test should be fed to MISR inputs that are separated by one or more feedback taps.

2.5.4. A simple BIST IC design

Figure 2.20 shows how a complete self-testing circuit can be constructed using an LFSR and a MISR. This is based on the use of scan-design.

One output of the LFSR feeds into the scan path, while the others are able to drive the circuit inputs during testing. (During normal operation of the chip, the outputs of the LFSR would be set to high impedance.) Similarly, one MISR input is fed from the output of the scan path, while the others are fed from the circuit outputs.

During test, operation of this circuit is as follows:

(1)   Signal the start of the test to the control circuit which will then reset the LFSR and MISR to their starting states.

(2)   Select the scan-test mode. Clock the complete circuit (including the LFSR) a sufficient number of times to fill the scan path with pseudo-random data. Together with the data applied from the LFSR to the circuit inputs, this forms the first test pattern for the combinational logic.

(3)   Enable the MISR. Turn the scan-test mode off and clock the circuit once. This causes the results of the test to be loaded into the scan path or, in the case of those at the circuit outputs, to be captured by the MISR.

(4)   Repeat step (2) keeping the MISR enabled. As the results of the first test are shifted out into the MISR, the second test stimulus is shifted in from the LFSR.

Continue to repeat stages (2) and (3) until a sufficient number of tests has been applied.

(5)   At the end of the test, inspect the contents of the MISR and compare to the expected fault-free value.

2.5.5. The BILBO

The **built-in logic block observer** (BILBO) (see Figure 2.21) combines the functions of an LFSR and a MISR into a single building block that can be used in a self-testing circuit design (Konemann *et al.*, 1979).

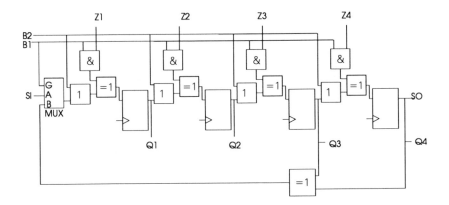

Figure 2.21 The BILBO.

Table 2.5 Operating modes of the BILBO.

B1	B2	Operation
1	1	The circuit behaves as a parallel-input, parallel-output register. Data presented at the inputs Z is loaded into the flip-flops on each clock cycle. This is the normal operating mode of the BILBO.
0	0	The circuit becomes a shift register. Data presented at the serial input (S1) appears at the serial output (S2) after an appropriate number of clocks has been applied. By using this mode together with the parallel register mode, scan testing is possible.
1	0	The circuit becomes a MISR. Data presented at the inputs Z is compressed into a signature that can be shifted out for examination by selecting the shift register mode. If the data inputs can be held constant, then a pseudo-random test sequence is generated at the parallel outputs Q.
0	1	The flip-flops are reset when the clock transition occurs.

The BILBO has four operating modes that are selected using inputs B1 and B2 as shown in Table 2.5.

A BILBO could be positioned between two combinational circuit blocks, for example as shown in Figure 2.22. During testing of block 1, the BILBO would be used as a MISR to generate a signature from the results of the test applied at the inputs to the block. During testing of block 2, the inputs to block 1 would be held constant, allowing the BILBO to generate a pseudo-random test for block 2.

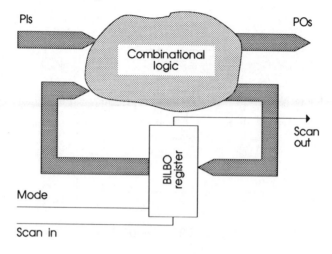

Figure 2.22 Use of the BILBO.

2.5.6. Macro test

While the use of LFSRs, signature analysers, MISRs, and BILBOs is an effective way of converting an arbitrary circuit into a self-testing design, the use of other techniques may result in a better solution for certain types of building block (macro) — for example, better quality test for lower added silicon. This is particularly true where the detail of the silicon layout and of the manufacturing process are known. Examples would be highly-regular macro designs, such as RAMs, read-only memories (ROMs), and **arithmetic logic units** (ALUs).

ASIC vendors are increasingly designing such macros with their own BIST capability — for example, to give a self-testing RAM. Generic versions of such macros are included in the ASIC vendor's library, allowing chip designers to request, say, a RAM of the size required for their particular circuit.

A number of self-testing macros might be included in a complete chip design and would need to be connected to a small built-in controller that would control their behaviour during test execution. Testing of the logic not contained in the self-testing macros is organized by the chip designer, and could be based on scan paths or LFSR-based self-test.

2.5.7. Board-level self-test

LFSRs and MISRs can also be used to allow the development of self-testing loaded boards. For example, a MISR could be connected so as to compress a data stream on a microprocessor bus into a signature.

More often, though, self-test is achieved by the provision of additional firmware on the board. This firmware is executed by the on-board processor when test is selected and allows the processor to act as the on-board ATE. The processor writes data into the various peripheral chips and reads their responses. A firmware equivalent of a MISR may be used to compress the result data into a compact signature.

Some guidelines for the design of self-testing boards are contained in Part 2.

References

Books

Readers who intend to design ASICs for use on their board designs should refer to the following books, which contain more detailed discussions of scan design, self-test, and boundary-scan (including ANSI/IEEE Std 1149.1).

Bardell P., McAnney W.H., and Savir J. (1987). *Built-in Test for VLSI — Pseudorandom Techniques*. New York: J. Wiley & Sons.

Bennetts R.G. (1984). *The Design of Testable Logic Circuits*. London: Addison-Wesley.

Maunder C.M. and Tulloss R.E. (1990). *The Standard Test Access Port and Boundary-Scan Architecture*. Los Alamitos, Calif.: IEEE Computer Society Press.

Papers

Ando H. (1980). Testing VLSI with random-access scan. *Proceedings IEEE Compcon,* Spring.

Eichelberger E.B. and Williams T.W. (1978). A logic design structure for LSI testability. *Journal of Design Automation and Fault Tolerant Computing,* (May), 2(2), 165-178.

Frohwerk R.A. (1977). Signature analysis: a new digital field service method. *Hewlett-Packard Journal,* 2-8.

Goel P. (1980). Test costs, analysis, and projections. In *Proc. IEEE Design Automation Conference,* Philadelphia, PA, 77-82.

Konemann B., Mucha J., and Zweihoff G. (1979). Built-in logic block observation techniques. In *Proc. IEEE Test Conference Proceedings,* Philadelphia, 37-41.

McCluskey E. (1984). A survey of design for testability scan techniques. *VLSI Design,* (December), 38/39/42/46/48/59/60/61.

Reinerstein D.G. (1983). Whodunit? The search for the new-product killers. *Electronic Business*, (July), 62/4/6.

CHAPTER 3.

Boundary-Scan

3.1. Introduction

Two continuing trends are having a significant adverse impact on the cost of testing loaded printed wiring boards (PWBs):

(1)   *Increasing complexity*: As integrated circuits (ICs) become more complex, the difficulty of generating a test for loaded boards increases. For in-circuit testing, the generation of a test module for a new IC design could take several man-months. For functional testing,

test generation times are significantly longer, due to the need to propagate test data through the complex ICs while applying tests to other chips on the board. Test lengths also increase as complexity rises, pushing up the cost of applying the finished test. In in-circuit testing, the maximum test length is constrained by a time limit imposed to ensure that components are not damaged by backdriving. Often, therefore, the result of increasing complexity is that less thorough testing is possible.

(2)   *Greater miniaturization:* Test techniques that became widely used during the 1980s — in-circuit, functional, cluster, and emulation testing — depend significantly on the ability to make contact with connections internal to the loaded board. Bed-of-nails or hand-held probes must be connected to such connections during test application and/or fault diagnosis. The use of surface-mount assembly techniques, particularly when coupled with double-sided component mounting and the use of buried vias to connect layers of interconnect on the PWB, reduces board geometries — making the finished product smaller and, unfortunately, more difficult to probe.

The aim of ANSI/IEEE Std 1149.1 (IEEE, 1990) — the *Standard Test Access Port and Boundary-Scan Architecture* — is to provide the basis of solutions to these problems. The key is the elimination of the need for physical probing of the loaded board, which is achieved by building an electronic test-access mechanism (the boundary-scan path) into the integrated circuits themselves.

This chapter provides a guide to the principal features defined by the standard and to their operation. It is intended as a prelude to the standard itself, not as a substitute for it. In particular, it is recommended that readers who intend to implement integrated circuits, design tools, or test systems that support the standard should read the standard document before doing so. Also, while this chapter shows how basic test operations can be performed, it is not a complete guide to the potential applications of the standard. For a view of the wide range of applications of the standard, the reader is directed to Maunder and Tulloss, 1990.

3.2. A chip-level view

Figure 3.1 gives an overall view of an IC design that conforms to ANSI/IEEE Std 1149.1. The circuitry can be broken down into two parts:

(1)   *The system logic.* This is the circuitry that performs the 'normal' function for which the chip was designed. For example, it could be the logic necessary to build a microprocessor, a communications interface, or a counter.

(2) *The test logic.* This is the circuitry defined by the standard and includes all the blocks in Figure 3.1 other than the system logic. The test logic is used during testing of the IC and of the board onto which the IC is assembled. It does not contribute to the normal operation of the system logic.

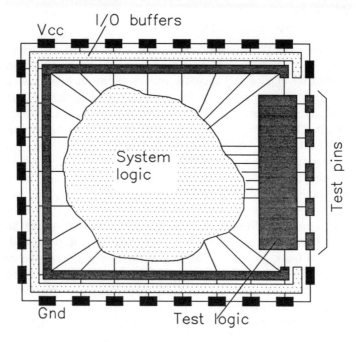

Figure 3.1 Chip-level view.

Some circuitry may be shared between the system and test logic. For example, a register in the system logic may have test modes of operation — perhaps it might be a BILBO or similar device (Konemann *et al.*, 1979). Under these circumstances, the circuitry is regarded as system logic while it contributes to the normal chip function and as test logic while it participates in test operations.

3.3. The test logic architecture

The top-level schematic of the test logic defined by ANSI/IEEE Std 1149.1 includes three key circuit blocks (Figure 3.2):

☐ *The TAP controller.* This responds to control sequences supplied through the test access port (TAP — see Section 3.4) and generates

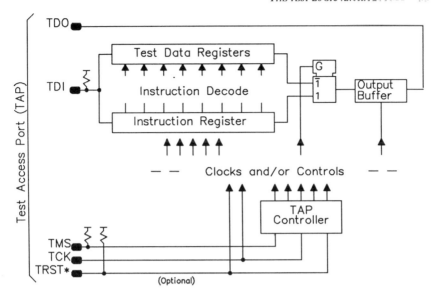

Figure 3.2 ANSI/IEEE Std 1149.1 test logic.

the clocks and control signals required for correct operation of the other circuit blocks.

❏ *The instruction register.* This shift register based circuit is serially loaded with the instruction that selects a test to be performed.

❏ *The test data register.* This is a bank of shift register based circuits (see Section 3.7). The stimuli or conditioning values required by a test are serially loaded into the test data register selected by the current instruction. Following execution of the test, results in test data registers can be shifted out for examination.

These circuit blocks are connected to a TAP which includes the four or, optionally, five signals used to control the operation of tests and to allow serial loading and unloading of instructions and test data. The TAP on an IC is directly analogous to the 'diagnostic' socket provided on many automobiles — it allows an external test processor to control and to communicate with the various test features built into the product. When a number of ICs that implement the standard are combined on a PWB, they can be arranged in a single daisy chain with:

❏ the test data input (TDI) input of the first IC in the chain connected to the board edge or to the output of an on-board master device;

❏ subsequent ICs each having their TDI connected to their predecessor's test data output (TDO); and

❏ the last IC's TDO being fed to the board edge or to the input of the on-board master device.

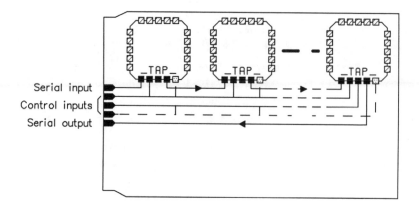

Figure 3.3 Board-level connection of standard ICs.

This arrangement, shown in Figure 3.3, supports the serial transfer of test data and instructions. Other arrangements are possible.

3.4. The TAP

The TAP contains four or, optionally, five pins. These are:

❏   *The test clock input (TCK).* This is independent of the system clock(s) for the chip so that test operations can be synchronized between the various chips on a PWB.

❏   *The test mode select input (TMS).* The operation of the test logic is controlled by the sequence of 1s and 0s applied at this input. The sequence on TMS directs the TAP controller in its generation of the clock and control signals required by the other test logic blocks.

❏   *The test data input (TDI).* Data applied at this serial input is fed either into the instruction register or into a test data register, depending on the sequence previously applied at TMS.

❏   *The test data output (TDO).* This serial output from the test logic is fed either from the instruction register or from a test data register depending on the sequence previously applied at TMS. During shifting, data applied at TDI will appear at TDO after a number of cycles of TCK determined by the length of the register included in the serial path. When data is not being shifted through the chip, TDO is set to an inactive drive state (for example, high impedance).

❏   *The optional test reset input (TRST*).* The test logic is designed so that it can be reset synchronously under control of TCK and TMS. TRST* provides a supplementary reset mechanism. The test logic is reset when a 0 is applied at TRST*.

The TDI, TMS, and TRST* inputs are either equipped with a pull-up resistor or otherwise designed such that, when they are not driven from an external source, the test logic perceives a logic 1.

3.5. The TAP controller

A key goal during the development of ANSI/IEEE Std 1149.1 was to keep the number of pins in the TAP to a minimum, based on the knowledge that many ICs are pin- (rather than silicon-) limited. As test engineers are only too aware, chip designers are always reluctant to allocate pins for test purposes.

The TAP controller allows this goal to be met. It is a 16-state finite state machine that operates according to the state diagram shown in Figure 3.4. Note that in the states whose names end '-DR' the test data registers operate, while in those whose names end '-IR' the instruction register operates. A move along a state transition arc occurs on every rising edge of TCK. The 0s and 1s shown adjacent to the state transition arcs show the value that must be present on TMS at the time of the next rising edge of TCK for the particular transition to occur.

Eight of the 16 controller states determine operation of the test logic, allowing the following test functions to be performed:

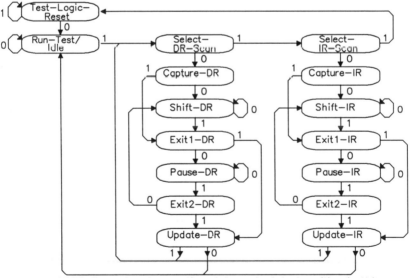

NOTE: The value shown adjacent to each state transition in this figure represents the signal present at TMS at the time of a rising edge at TCK.

Figure 3.4 State diagram for the TAP controller.

❏   *Test-Logic-Reset*. All test logic is reset, allowing normal operation of the chip to occur without interference. Regardless of the starting state of the TAP controller, the *Test-Logic-Reset* controller state is reached by holding the TMS input at 1 and applying five rising edges at TCK. Alternatively, where TRST* is provided, it can be used to force the controller asynchronously into the *Test-Logic-Reset* controller state at any desired point during circuit operation.

❏   *Run-Test/Idle*. The operation of the test logic in this controller state depends on the instruction held in the instruction register. When the instruction is, for example, one that activates a self-test, then the self-test will be run in this state. If, in another case, the instruction in the instruction register is one that selects a data register for scanning, then the test logic will be idle.

❏   *Capture-DR*. Each instruction must identify one or more test data registers that are enabled to operate in test mode when the instruction is selected. In this controller state, data is loaded from the parallel inputs of the selected test data registers into its shift register paths.

❏   *Shift-DR*. Each instruction must identify a single test data register that is to be used to shift data between TDI and TDO in the *Shift-DR* controller state. Shifting allows the previously captured data to be examined and new test input data to be entered.

❏   *Update-DR*. This controller state marks the completion of the shifting process. Some test data registers may be provided with a latched parallel output to prevent signals applied to the system logic, or through the chip's system pins, from rippling while new data is shifted into the register. Where such test data registers are selected by the current instruction, the new data is transferred to its parallel outputs in this controller state.

❏   *Capture-IR, Shift-IR, and Update-IR*. These controller states are analogous to *Capture-DR, Shift-DR,* and *Update-DR* respectively but cause operation of the instruction register. By entering these states, a new instruction can be entered and applied to the test data registers and/or other specialized circuitry. The new instruction becomes 'current' in the *Update-IR* controller state.

Note that in the *Update-DR* and *Update-IR* controller states, the described action takes place on the falling edge of TCK. In all other states, the described action takes place on the rising edge of TCK just before the controller leaves the state (see Figure 3.5). Also note that TDO is active only in the *Shift-DR* and *Shift-IR* controller states.

    In the remaining eight controller states, no operation of the test logic occurs — that is, the test logic is 'idle.' The 'pause' states (*Pause-DR* and *Pause-IR*) are provided to allow the shifting process to be temporarily halted. This might occur while an ATE or other equipment controlling the test logic fetches more test data from backup memory (for example, disc).

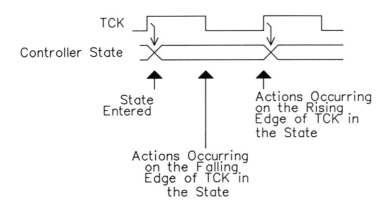

Figure 3.5 The timing of events within a controller state.

The final six controller states (*Select-DR-Scan, Select-IR-Scan, Exit1-DR, Exit1-IR, Exit2-DR*, and *Exit2-IR*) are decision points that allow choices to be made as to the route to be followed around the controller's state diagram. For example, in the *Exit1-DR* controller state a choice is made between entry into the *Pause-DR* state or entry into the *Update-DR* state.

3.6. The instruction register

The instruction register provides one of the alternate serial paths between TDI and TDO. It operates when the instruction scanning portion of the controller state diagram is entered (that is, the portion where state names end '-IR').

The instruction register allows test instructions to be entered into each IC along the board-level path. At the board level, the instruction registers are daisy-chained together in the *Shift-IR* controller state (Figure 3.6), so a different instruction can be loaded into each chip on the path if required.

The instruction register is a parallel-in, parallel-out shift register. The parallel output is latched so that a new instruction can be shifted in without altering the instruction applied to the remainder of the test logic. The latched output is updated from the shift register path in the *Update-IR* controller state; at this time, the new instruction becomes 'current'. In the *Test-Logic-Reset* controller state, the latched output is reset — to either the IDCODE or the BYPASS instruction depending on the set of test data registers built into the particular IC (see Sections 3.8 and 3.9).

The instruction register must contain at least two stages (shown cross-hatched in Figure 3.7). No maximum length is defined, since this will be determined by the number of test instructions provided by the particular IC. Stages I1 and I0 (that is, the stages located nearest to the serial output)

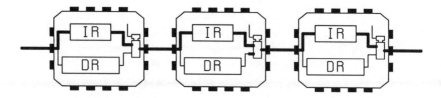

Figure 3.6 Daisy-chain connection of instruction registers.

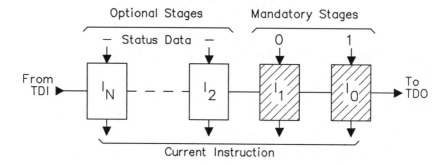

Figure 3.7 The instruction register.

must be set to 0 and 1 respectively in the *Capture-IR* controller state. These fixed values assist in detecting and locating faults in the serial path through chips on a board (Maunder and Tulloss, 1990).

Instruction register stages numbered I2 or greater are optional and can have a parallel input from which data (typically, status information) is loaded.

3.7. The test data registers

The test logic design provides for a bank of test data registers as shown in Figure 3.8.

ANSI/IEEE Std 1149.1 specifies the design of three test data registers, two of which must be included in the design. The mandatory test data registers are the bypass and boundary-scan registers. The provision of a device identification register is optional and further design-specific test data registers can be added as appropriate to a given design.

All test data registers are shift register based and operate according to the same principles:

❏   Operation of the various test data registers is controlled according to the current instruction. An instruction can place several test data registers into their test mode of operation, but it can select only one register for connection as the serial path between TDI and TDO in the *Shift-DR* controller state. On the other hand, it is important to note that one or more physical registers can be configured as one (virtual) test data register by a given instruction.

❐   Registers that are not enabled for test operation by the current instruction are configured so that they do not interfere with operation of the on-chip system logic (in the case of a register that can operate in either system or test mode, the system mode will be selected).

❐   Registers enabled for test operation by the current instruction will load data from their parallel inputs (if any) in the *Capture-DR* controller state and will make any new data available at their latched parallel outputs (if any) in the *Update-DR* controller state. In other words, the results of a test are sampled in the *Capture-DR* controller state and the new test stimulus is available, at the latest, in the *Update-DR* controller state.

❐   Where test execution is required between the *Update-DR* and *Capture-DR* controller states (for example, execution of a self-test), this occurs in the *Run-Test/Idle* state.

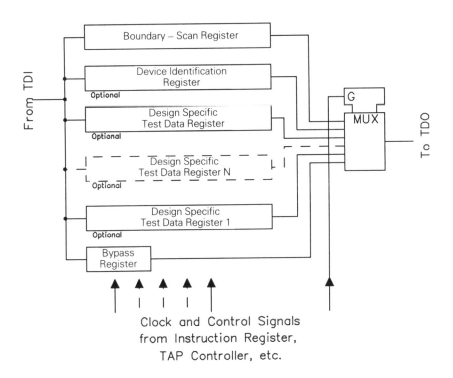

Figure 3.8 Test data registers.

☐   The register selected by the instruction to be the serial path between
TDI and TDO will shift data from TDI towards TDO in the *Shift-DR*
controller state. All other test data registers enabled for test operation
will hold their state while shifting occurs.

3.8. The BYPASS instruction

Every IC that conforms to the standard must support the BYPASS
instruction. A value (but not necessarily the only value) for the BYPASS
instruction must be 'all-1s' (that is, a logic 1 entered into each stage of the
instruction register). In ICs that do not include the optional device
identification register, the BYPASS instruction is forced onto the instruction
register's output in the *Test-Logic-Reset* state and thus becomes the 'current'
instruction whenever the test logic is reset.

The BYPASS instruction selects the bypass register as the serial path
between TDI and TDO during the *Shift-DR* controller state. This register
consists of a single parallel-in, serial-out shift register stage that loads a
constant logic 0 in the *Capture-DR* controller state when the BYPASS
instruction is selected. The bypass register does not have a parallel data
output so there is no significance to the data present in the register when
shifting is completed. The operation of the register cannot interfere with that
of the on-chip system logic.

As an example of an occasion when the bypass register might be
used, consider a board containing 100 ICs, all with boundary-scan and
connected into a single serial chain, a small part of which is shown in Figure
3.9. Assume that a need arises to access a test data register located in IC57,
but that it is desired not to interfere with the operation of the remaining 99
ICs. (An example of such a situation might be when the target chip includes
a 'shadow' test data register that permits the state of its key internal registers
to be read.)

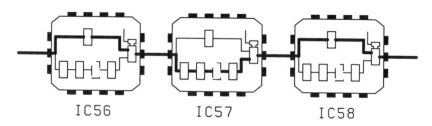

IC56   IC57   IC58

Figure 3.9   Use of the bypass register.

In this case, the required instruction would be loaded into IC57, with the BYPASS instruction being loaded into the other ICs. The serial bit stream shifted into TDI during the instruction scanning cycle would be:

$$111.......1111CCC...CCC1111.........111$$

where CCC...CCC is the instruction to be loaded into IC57. As a result of use of the *all-1s* value for the BYPASS instruction, the complexity of the bit stream input to the serial path is considerably reduced. This is an important consideration, since it reduces the data storage requirement for the automatic test equipment (ATE) or bus master chip that controls the operation of the board during test.

Once the instructions are loaded, a minimum length serial path to and from the target chip is set up that allows access to the chip of interest in the minimum possible time, increasing test throughput.

3.9. The IDCODE and USERCODE instructions

The IDCODE and USERCODE instructions select use of the optional device identification register. In every IC that includes the device identification register, the IDCODE instruction is forced onto the instruction register's output in the *Test-Logic-Reset* state and thus becomes the 'current' instruction whenever the test logic is reset.

The device identification register allows a binary data pattern to be read from the chip that identifies the manufacturer, the part number, the variant, and (where appropriate) the programmed state. This information might be used to:

❏   adjust test program execution, depending on the source and/or variant of each chip present on the board;

❏   verify that the correct IC (or correctly-programmed IC) has been mounted in each board location; or

❏   establish which member of a plug-compatible family of boards is being tested.

The register contains 32 parallel-in, serial-out shift register stages. Like the bypass register, the device identification register does not have a parallel output and, in consequence, there is no significance to the data contained in the register when shifting terminates. Also like the bypass register, operation of the device identification register can occur without interfering with normal chip operation.

Where an IC is programmed off-line (for example, by blowing fuses or through some other non-reversible process), each stage must have a pair

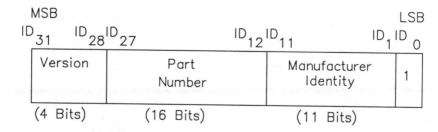

Figure 3.10 Structure of the device identity code.

of alternative data inputs so that two different 32-bit codes can be loaded —
one to identify the device and one to identify its programming. The former is
loaded when the IDCODE instruction is selected, while the latter is loaded
when the USERCODE instruction is selected. In all other types of IC, only
one data input is required and the USERCODE instruction is not provided.

The structure of the data loaded into the device identification register
in response to the IDCODE instruction is shown in Figure 3.10. There are
four separate fields:

(1) *The header.* ID0 loads a constant logic 1.

(2) *The manufacturer code.* ID11-ID1 load an 11-bit manufacturer code.
This code is derived from a scheme managed by the **Joint Electron
Device Engineering Council** (JEDEC) — see IEEE (1990) and
JEDEC (1986). The code can uniquely identify up to 2032
manufacturers (since 16 codes are not used). If more than 2032
manufacturer codes are issued by JEDEC, then the scheme will result
in the reuse of some code values within the manufacturer code field.
However, the chance that a component from an incorrect
manufacturer will have the same code and the same test functionality
is acceptably low.

(3) *The part number code.* ID27-ID12 provide a 16-bit part number,
chosen by the manufacturer to distinguish a chip from the others that
the company sells. In cases where more than 2^{16} chip types are
offered by a manufacturer, part number codes might have to be
reused. The objective is to minimize the chance that an incorrect chip
in a given position on a board will have the same part number as the
correct chip type — not to provide absolute identification of the IC.

(4) *The version number code.* For chips that are manufactured in several
different versions through their lives, bits ID31-ID28 can be used to
distinguish up to 16 variants.

The data loaded by the USERCODE instruction may be organized as the
part's programmer sees fit. It must be programmable at the same time (and in
the same way) as the function of the chip is programmed.

3.10. Boundary-scan register instructions

The boundary-scan register is a shift register based structure comprising a variety of different cell designs matched onto the requirements of the particular component. Different cell designs are used according to the type of system pin concerned (input, output, 3-state, bidirectional) and according to the set of boundary-scan instructions supported.

A simplified view of a boundary-scan register is shown in Figure 3.11.

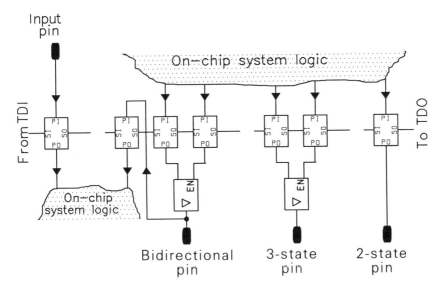

Figure 3.11 A simplified view of the boundary-scan register.

An example implementation for a cell that could be used in each of the locations shown in Figure 3.11 is given in Figure 3.12.

The connections labelled PI, PO, SI, and SO in Figure 3.12 are connected to adjacent cells, the on-chip system logic, and the system pins as shown in Figure 3.11. In Figure 3.12, the signals ClockDR, ShiftDR, and UpdateDR are generated by the TAP controller in response to changes at the TCK and TMS input pins. The Mode input is controlled according to the type of pin connected to the cell (input, output, etc.) and the specific instruction selected.

Use of this sample cell design, with appropriate signals supplied to the Mode input of each cell, will result in a component that supports the SAMPLE/PRELOAD, EXTEST, and INTEST instructions described below. Other cell designs are possible that meet the requirements of this standard for different sets of instructions. For example, if the INTEST instruction were not supported in a given design, R2 and M2 would be omitted from

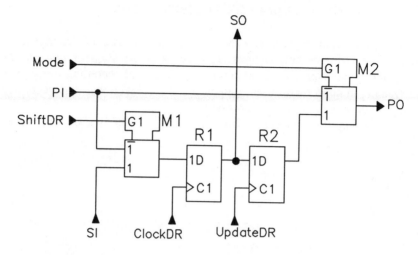

Figure 3.12 An example boundary-scan cell design.

cells that feed data from a system pin to the on-chip system logic. The input, PI, would then be connected directly through the cell to the output, PO.

As the boundary-scan register instructions are discussed, we note for each:

(1) whether the instruction is mandatory or optional;

(2) which test data registers can be connected in the serial path between TDI and TDO;

(3) restrictions (if any) on the choice of binary codes for the instruction; and

(4) flow of data between the component's system pins, the boundary-scan register cells, and the on-chip system logic.

3.10.1. SAMPLE/PRELOAD

This mandatory instruction allows a snapshot of the normal operation of the component to be taken and examined. The taking of this snapshot is arranged so as to have no effect on the system operation that it monitors. The instruction also allows data values to be loaded onto the latched parallel outputs of the boundary-scan shift register prior to selection of other boundary-scan test instructions. This instruction must select only the boundary-scan register to be connected between TDI and TDO in the Shift-DR controller state. There is no required binary value that must be decoded as the SAMPLE/PRELOAD instruction. When SAMPLE/PRELOAD is the current instruction, test logic operation is not permitted to have any effect on operation of system logic or on the flow of signals between the IC's system pins and on-chip system logic (that is, the Mode signals for all cells should be set to 0).

The sampling mechanism of the SAMPLE/PRELOAD instruction has the effect of loading the states of all signals flowing through system pins into their corresponding boundary-scan register cells on the rising edge of TCK during the *Capture-DR* controller state (Figure 3.13). (As in later figures, the paths followed by signals when the appropriate instruction is selected are shown as bold lines.) This capability allows the state of the interconnect network of a machine to be captured at a desired moment (for example, upon a failed parity check) by having the requisite edge of the signal feeding TCK triggered by a specified system event.

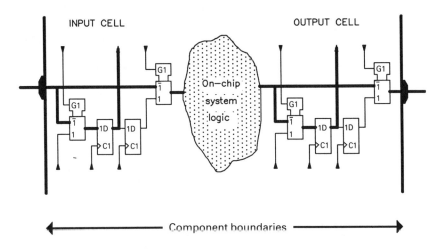

Figure 3.13   Signal flow when the SAMPLE/PRELOAD instruction is selected.

The preloading mechanism allows an initial data pattern to be placed at the latched parallel outputs of boundary-scan register cells (for example, in cells connected to pins driving off-chip) prior to selection of some other boundary-scan test operation. For example, the EXTEST instruction (to be discussed in Section 3.10.2) is used for testing the interconnection of chips. It is very desirable that predetermined and non-damaging signals be driven out of chips on a board while the first test pattern to be applied to the interconnect is being scanned into the boundary-scan register. This aim can be achieved by preloading the same boundary-scan register. As soon as the EXTEST instruction has been transferred to the parallel output of the instruction register, preloaded data will be driven through the system output pins of each IC; this situation will remain in effect while scanning of external test patterns is going on.

Shifting of data for sampling and preloading can occur concurrently when required — while sampled data is shifted out, preloading of new data can occur.

3.10.2. EXTEST

One of the principal motivations for the development of the standard was the need for a non-contact method of testing board (and system) interconnect — see Beenker (1985), Jarwala and Yau (1989), and Yau and Jarwala (1989). The EXTEST instruction is the key to the standard's response to that need. The EXTEST instruction is, therefore, mandatory. There may be more than one binary code that is interpreted as EXTEST; however, one of these codes must be the 'all-0's' instruction code.

When EXTEST is the current instruction, the boundary-scan register is the one and only register that is to be connected between TDI and TDO for data scanning purposes. While the EXTEST instruction is selected in a chip:

(1)   system logic of the chip must be controlled such that it cannot be damaged as a result of signals received at the system input or system clock input pins;

(2)   the state of all signals driven from system output pins is completely defined by the data shifted into the boundary-scan register and changes only on the falling edge of TCK in the *Update-DR* controller state; and

(3)   the state of all signals received at system input pins is loaded into the boundary-scan register on the rising edge of TCK in the *Capture-DR* controller state.

The EXTEST instruction allows circuitry external to the component package to be tested. Typically such circuitry would be the board interconnect. Clusters of components that lack boundary-scan registers can be statically tested using the same functionality although complex clusters would require more sophisticated diagnostic systems than would interconnect alone. During use of the EXTEST instruction, boundary-scan register cells at output pins are used to apply test stimuli, while those at input pins capture test results (Figure 3.14). Captured results are scanned out of the serially linked boundary-scan registers of a board while the next set of test input values is scanned in.

As was suggested in Section 3.10.1, the first test stimulus to be applied using the EXTEST instruction should be shifted into the boundary-scan register using the SAMPLE/PRELOAD instruction. This is the most judicious approach because, when the change to the EXTEST instruction takes place in the next occurring *Update-IR* controller state, known data will

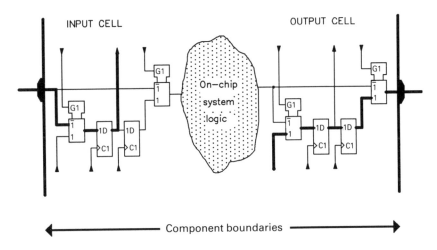

Figure 3.14 Signal flow when the EXTEST instruction is selected.

be driven immediately from the component onto its external connections. Where a total of N tests are to be applied using the EXTEST instruction, stimuli for tests 2 to N will be shifted in while the results from tests 1 to N - 1 are shifted out. Note that, while the results from the final test (test N) are shifted out, a determinate set of data should be shifted in that will leave the board in a consistent state at the end of the shifting process. This can be achieved by shifting the stimuli for test N (or indeed any other test) into the boundary-scan register for a second time.

Note that the boundary-scan register cells located at input pins may optionally be designed to allow signals to be driven into the on-chip system logic when the EXTEST instruction is selected. This allows user-defined values to be established at the system logic inputs, preventing misoperation in response to noise signals arriving from the board-level interconnect. The values driven may either be constant for the duration that EXTEST is selected (for example, by including a blocking gate at the input to the system logic) or they may be loaded serially through the boundary-scan register.

The EXTEST instruction can be entered by holding TDI at a constant low value and completing an instruction-scan cycle of sufficient duration to fill each instruction register on the board-level serial data path. As in the case of the BYPASS instruction, this may simplify demands on the ATE or bus master device which controls a test.

The data loaded into boundary-scan register cells located at system output pins (2-state, 3-state, or bidirectional) in the *Capture-DR* controller state when the EXTEST instruction is selected should be independent of the operation of the on-chip system logic. Where followed, this recommendation ensures that data shifted out of the component in response to the EXTEST instruction is not altered by the presence of faults in a chip's system logic. This simplifies diagnosis since any errors in the output bit stream can only be caused by faults in off-chip circuitry, in board-level interconnections, or in boundary-scan registers used to apply the test.

3.10.3. INTEST

INTEST is one of two optional instructions defined by the standard that allow testing of on-chip system logic after a component is assembled on a board. The binary value(s) that are decoded as the INTEST instruction may be selected by the component designer's specification. Using the INTEST instruction, test stimuli are shifted in one at a time and applied to the on-chip system logic. To achieve this, on-chip system logic must be capable of single-step operation while INTEST is the current instruction. Internal test results are captured in the the IC's boundary-scan register and are examined by subsequent shifting. Initial data set-up for such a test can be achieved using the SAMPLE/PRELOAD instruction.

The INTEST instruction must select the boundary-scan register to be the one and only register connected between TDI and TDO for shifting access. When INTEST is the current instruction:

(1)   the state of all signals driven from system output pins must be completely defined by data previously shifted into the boundary-scan register and is allowed to change only on the falling edge of TCK in the *Update-DR* controller state;

(2)   the state of all non-clock signals driven into the system logic from the boundary-scan register must be completely defined by data previously shifted into the register;

(3)   the state of all signals output from the system logic to the boundary-scan register must be loaded into the register on the rising edge of TCK in the *Capture-DR* controller state.

The INTEST instruction allows slow speed (static) testing of on-chip system logic (Figure 3.15). Each test pattern and response must be shifted through the boundary-scan register. While an INTEST-based test is proceeding, the logic values at the component output pins are defined from the boundary-scan register. This requirement ensures that surrounding components on an assembled board are supplied known signal levels while on-chip system logic testing is in progress. A consistent, 'safe' set of data values would be shifted into the appropriate stages of the boundary-scan register using the SAMPLE/PRELOAD instruction prior to selection of the INTEST instruction. This 'safe' data pattern is then reloaded into boundary-scan control cells and the cells associated with chip outputs each time a new INTEST test pattern is shifted into the boundary-scan register.

As noted above, the approach taken with the INTEST instruction requires that on-chip system logic can be operated in a single step mode —

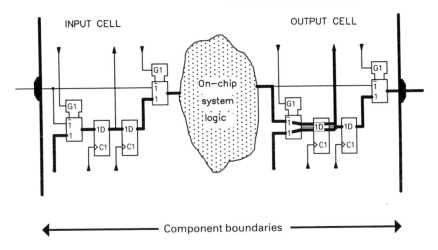

Figure 3.15   Signal flow when the INTEST instruction is selected.

where the circuitry moves one step forward in its operation each time shifting of the boundary-scan register is completed. Note that, for each such test step, the latched parallel output of the boundary-scan cell at the system output pin is updated from data shifted in before the contents of the shift register is overwritten with a new test response.

While the INTEST instruction is selected, an IC's boundary-scan register assumes the role of the fixture and pin electronics of an ATE used for stand-alone component testing. Cells at non-clock system input pins are used to apply test stimuli, while those at system output pins capture responses. Stimuli and responses are moved into and out of the circuit by shifting the boundary-scan register.

To achieve single step operation, on-chip system logic can be expected to receive a sequence of clock events between application of the stimulus and capture of the response. A designer's specification of boundary-scan cells for system clock input pins would allow clocks for on-chip system logic to be provided in several ways while INTEST is the current instruction. Here are some examples:

(1)   The signals received at system clock pins can be fed directly to the on-chip system logic as they would be during non-test operation. When this approach is taken, off-chip clock sources should be handled in such a way that, during internal testing via INTEST, clock signals received by the component change state only in the *Run-Test/Idle* controller state. In this way, on-chip system logic operation can be inhibited while test data is shifted through the boundary-scan register. Figure 3.16 illustrates how a system clock applied to a component should be controlled during INTEST-based testing of on-chip system logic.

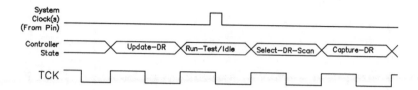

Figure 3.16 Control of applied system clock during INTEST.

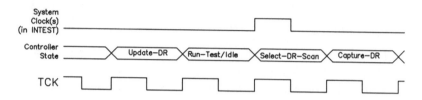

Figure 3.17 Use of TCK as clock for on-chip system logic during INTEST.

(2)   On-chip system logic can be supplied with clock signals derived from the input signal to TCK in the *Run-Test/Idle* controller state. In all other controller states, such clocks should not change state (Figure 3.17).

(3)   Circuitry may be built into the component which allows on-chip system logic to complete one step of operation upon entry into the *Run-Test/Idle* controller state. For example, if the component were a microprocessor, it would be permitted to complete a single processing cycle; this might be achieved by internal generation of a pulse on the processor's hold signal. In such a case, the clock(s) applied at system clock pin(s) during the test could be free-running.

(4)   Clock signals can be loaded serially via the boundary-scan path in the same manner in which non-clock signals for the on-chip system logic are supplied. This would require the boundary-scan register to be loaded for each distinct clock signal state — twice for a single-phase clock. This may be a hazard-prone operation for certain circuit designs.

    The standard recommends that for boundary-scan register cells located at system input pins (clock or non-clock) or at bidirectional pins configured as inputs, the data loaded in the *Capture-DR* controller state when the INTEST or RUNBIST (see Section 3.10.4) instruction is selected should be independent of the operation of off-chip circuitry or board-level interconnections. Where followed, this approach ensures that data shifted out of a component in response to the INTEST instruction is not altered by the presence of faults in off-chip system logic, board-level interconnections, and so on. This will simplify diagnosis; any errors in the output bit-stream can be caused only by faults in on-chip system logic or the boundary-scan register.

3.10.4. RUNBIST

In many cases, it may be desirable to put more of the burden of test into the product itself, rather than on an external tester. Contemporary built-in self-test (BIST) technology is beginning to make this possible with relatively little circuit overhead or performance penalty (Scholtz *et al.*, 1988). The optional RUNBIST instruction causes execution of a self-contained self-test of the component. Its binary encoding is left to the discretion of the designer. Use of the RUNBIST instruction allows a component user to determine the health of a component without the need to load complex data patterns and without the need for single step operation (as required for the INTEST instruction). With this capability in place, it becomes possible for all components on a board that offer the RUNBIST instruction to execute their self-tests concurrently or in groups limited by power consumption or heat dissipation requirements; thus, a rapid health check for assembled boards can be provided. This has particular relevance to field service and maintenance.

RUNBIST is an important instance of the flexibility that the standard offers in providing a gateway (the TAP) through which powerful instructions can be passed to an IC when it has become a component of an assembled system.

The standard allows the development of public and private instructions which could cause execution of BIST in various subsections of an IC. (Such capabilities could contribute to support of failure mode analysis.) The RUNBIST instruction is intended to serve as a model for other BIST instructions and, when it is feasible to do so, to be a vehicle by means of which designers could link all self-testing circuitry within an IC into a single BIST process for the entire entity. The standard recommends that RUNBIST be implemented wherever possible.

Self-test operation accessed via the RUNBIST instruction must execute only in the *Run-Test/Idle* controller state; and when RUNBIST is the current instruction, the test data register into which the results of the self-test(s) will be loaded must be connected for serial access between TDI and TDO in the *Shift-DR* controller state. (This register can be the boundary-scan register.) The result of the self-test(s) executed in response to the RUNBIST instruction must be loaded into the designated test data register no later than the rising edge of signal input to TCK in the *Capture-DR* controller state.

The developers of the standard intended the RUNBIST instruction to be an encouragement for designers to create as simple an interface as possible for self-testing ICs embedded in systems. Therefore, they required that where a test data register (other than the boundary-scan register) must be initialized prior to execution of the self-test, this must occur at the start of the self-test without any requirement to shift data into the component — there cannot be any requirement to enter seed values into any test data register other than the boundary-scan register. Moreover, where a component includes multiple self-test functions, these may be executed in response to the RUNBIST instruction either concurrently or in a sequence determined by the component manufacturer. In the latter case, all sequencing must be taken care of within the component without requiring alteration of instruction register contents or other external direction.

A number of the requirements in the standard are there for purposes of isolation — to assure a repeatable self-test result for fault-free circuits and to protect off-chip circuitry during self-test. The design of the component must ensure that results of self-tests executed in response to the RUNBIST instruction are not affected by signals received at non-clock system input pins; and data shifted out of a component, following the completion of an execution of a self-test accessed using the RUNBIST instruction, must be independent of the operation of off-chip circuitry or board-level interconnections.

When RUNBIST is the current instruction, the state of all signals driven from system output pins must be completely defined by data previously shifted into the boundary-scan register (for example, by use of the SAMPLE/PRELOAD instruction as described in Section 3.10.3); and states of parallel output registers or latches in boundary-scan register cells located at system output pins (2-state, 3-state, or bidirectional) are not permitted to change while the RUNBIST instruction is selected. In contrast to the INTEST instruction, the data values driven through the system output pins are held while the RUNBIST instruction is selected. For a boundary-scan register cell located at a system output pin (see Figure 3.12), the UpdateDR signal should be held at 0 while the RUNBIST instruction is selected and the Mode input should be held at 1.

While the RUNBIST instruction is selected, boundary-scan register cells associated with non-clock system inputs of a chip may simply be loaded with constant, 'safe' values. Alternatively, the boundary-scan register may act as a pattern generator and/or signature compactor in the *Run-Test/Idle* controller state — provided the states of parallel output registers or latches are unchanging as required above.

The specification of boundary-scan cells for system clock input pins allows the clocks for the on-chip system logic to be obtained in one of two ways while the RUNBIST instruction is selected:

(1)   The signals received at system clock pins can be fed directly to the on-chip system logic as they would be during non-test operation of the component. Where this is done, the design of the component must ensure that self-test executes only in the *Run-Test/Idle* controller state. However, the clock may be active in other controller states.

(2)   The on-chip system logic can be supplied with clock signals derived from the signal input at TCK in the *Run-Test/Idle* controller state. In this instance, in all other controller states, the clocks should not change state.

There are certain, additional, commonsense requirements placed on an IC that implements RUNBIST and on such an IC's documentation. All stages of the test data register selected by the RUNBIST instruction must be set to determinate logic states (0 or 1) in the *Capture-DR* controller state (that is, while the test result is loaded). A duration for the test executed in response to the RUNBIST instruction must be specified (for example, by citing a number of rising edges of the signal at TCK or of a system clock). Because it is likely that self-test in one component may complete before the self-test in another, the test results produced by the execution of the RUNBIST instruction and deposited in the specified test data register ready for shifting out of the IC must be stable through the period of delay before the *Capture-DR* controller state is entered, which period is likely to be unpredictable at the time of the component's design. Use of the RUNBIST instruction must give the same result in all versions of a component. These requirements were included in the standard to ensure that the test for an assembled board is independent of the versions of components mounted on it. Such independence is an important consideration when working in a maintenance or repair environment, where the versions of components used on a board may very well be unknown. The standard's requirement can be met by forming the exclusive-OR of the result from execution of the RUNBIST instruction with a fixed (version-dependent) pattern. The output from this function would become the result loaded into the boundary-scan register or other test data register connected between TDI and TDO during the time the RUNBIST instruction is current.

3.11. Machine-readable descriptions of ANSI/IEEE Std 1149.1-compatible ICs

While ANSI/IEEE Std 1149.1 defines the behaviour of the boundary-scan register and the other functional blocks shown in Figure 3.2, a number of parameters will vary from design to design. These include:

❏   the lengths of the instruction register and boundary-scan register (that is, the number of shift register based stages in each);

❏   the binary values used to encode each instruction defined by the standard (for example, EXTEST);

❏   the mapping of input/output pins onto boundary-scan cells; and

❏   the package pins assigned to the signals of the TAP.

A machine-readable language has been proposed that will facilitate communication of such parameters from the IC designer to the board designer, to **electronic design automation** (EDA) tools, or to a test system (Parker and Oresjo, 1991). In effect, the language — known as the **boundary-scan description language** (BSDL) — will act as an electronic data sheet for the test circuitry defined by the standard.

The language is a subset of the **VHSIC hardware description language** (VHDL) (IEEE, 1987) and is in the process of being made a formal part of ANSI/IEEE Std 1149.1. As will be discussed later, the availability of such a language will ease the development of EDA tools to support the use of ICs that are compatible with the standard.

3.12. Using boundary-scan

The boundary-scan paths included in ICs that conform to the standard will be used in many different ways by different companies. Key influences in determining precisely how the boundary-scan paths are used will include:

❐ the density of the board design;

❐ the type of loaded-board test system to be used;

❐ and, most significantly, the extent to which the board is populated with boundary-scan-compatible ICs (rather that ICs that do not conform to the standard).

Inevitably, many ICs that do not contain the test features defined by the standard will continue to be used. While some board designs may only use ICs that conform to the standard, others may use one or just a few.

In the following discussion, a range of situations illustrative of those that will occur in practice will be discussed. For each, a method of using the available boundary-scan circuitry to reduce test costs will be described and the impact on the board design and on EDA tools considered. The focus is on the move from in-circuit testing — the dominant test technique for loaded-boards in the 1980s — towards 100% boundary-scan testing.

Boundary-scan can also be used to ease functional (from the connector) testing where this technique is used. Here, the benefit of boundary-scan is that connections internal to the loaded board become controllable and/or observable without the need for probes or other forms of access. The greater the number of ICs with boundary-scan, the greater the controllability, observability, and testability of the loaded board. The particular advantage of boundary-scan in this context is that it will most likely be implemented in new state-of-the-art, high complexity ICs. These are precisely the components that cause the largest problems during test development for loaded boards.

3.12.1. Case 1: One or more isolated boundary-scan ICs

It will, of course, be some time before the inclusion of the features defined
by the standard is the norm, rather than the exception, in off-the-shelf
catalogue ICs. Therefore, it is probable that many boards will be designed
that contain either a single IC with boundary-scan or a small number of such
ICs that are isolated from one another by ICs without boundary-scan (that is,
there are no direct interconnections between the ICs with boundary-scan).
This situation is illustrated in Figure 3.18.

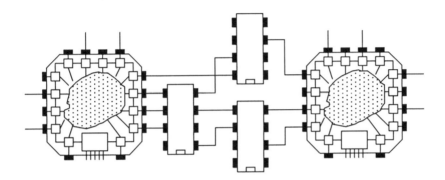

Figure 3.18 Isolated boundary-scan ICs.

In such situations, it is likely that an in-circuit test approach will be
used to test all ICs — those with and those without boundary-scan.
Therefore, bed-of-nails probes must be connected to all chip-to-chip
connections and to the TAP pins of the ICs with boundary-scan. The
requirement for probe access must be considered during layout of the PWB,
because there are limitations on the size of targets that can be probed and on
the proximity of one probe target to another. Further, it is a common
requirement that all probe targets must be located on one side of the PWB.
Unfortunately, few EDA vendors offer tools to help in this task.

In-circuit testing

In the case of in-circuit testing, each board interconnection is accessible to
the test system, typically through a bed-of-nails probe. To test that each IC
has been correctly soldered onto the board, the test system applies a test
through the probes connected to the IC's input and output points. This test
will check that:

(1)   1s and 0s driven onto board-level connections that feed input pins
       flow through the solder joint and into the IC; and
(2)   the outputs of the IC are able to drive connections to both 1 and 0.

For an IC without boundary-scan, these tasks are achieved by developing a test that detects (as a minimum) faults on the input and output pins. Task (1) is achieved by setting each input to both 1 and 0 and controlling the IC such that a change in the input signal (to 0 or 1, respectively) will cause an observable change at one or more output pins. Task (2) requires that each output is set to each logic value at some point during the test — a task that may be completed as a result of task (1).

Development of the test requires an understanding of the IC's normal behaviour, and therefore becomes more difficult as the complexity of the IC increases. For example, it has been estimated that some 6 man-months of effort would be required to develop a test module for the Motorola MC68040 microprocessor.

Two other factors should be noted:

(1)   Typically, the test module placed in an in-circuit tester's library will be developed on the assumption that no input pin is connected directly to power or ground and that all pins can be accessed by a bed-of-nails probe. Where this is not the case, a new test module will need to be developed. For complex ICs, this task may be costly.

(2)   The test will fail if there is a fault in the connection between any bed-of-nails probe and the IC. However, it may not allow the fault to be diagnosed, for example, to an open-circuit joint at a specific pin.

Developing in-circuit tests for boundary-scan ICs

If in-circuit testing is to be used, the principal benefit that may be gained from the existence of boundary-scan in an IC is a reduction in test development time, and thence in time-to-market. This benefit may be significant, since time-to-market is recognized as a key factor in determining the profitability of a new product.

For an IC with boundary-scan, an in-circuit test module can be created without any knowledge of the IC's 'normal' behaviour — all that is required is a specification of the boundary-scan path. The reason is that the test need only check that signals can flow between each bed-of-nails probe and the corresponding boundary-scan cell. Signals applied to the inputs of the IC can be observed using the 'load-and-shift' operation of the boundary-scan path and no longer need to be made observable at the outputs. Outputs can easily be set to both 1 and 0 using the boundary-scan path.

Given a knowledge of the boundary-scan path's behaviour (which is defined by the standard) and of the parameters of the test circuitry of the specific IC (which are defined by its BSDL file), an EDA tool could develop an in-circuit test module in a matter of minutes. Further, the tool could readily be rerun were any IC pins to be tied directly to power or ground on a particular board.

Because faults at input pins would cause an error to be detected by a specific boundary-scan cell, rather than at an output pin, the resulting test would allow location of faults to a particular probe-to-chip connection. (This is an improvement compared to a conventional in-circuit test.)

Design requirements

In addition to the continuing need for tools to assist in the layout of PWBs such that they can be reliably probed, a need will arise for tools to link the TAPs of the individual boundary-scan ICs into one or more board-level paths. The number and structure of such paths may need to be determined according to the ICs used in the particular design, as was discussed earlier.

The use of these tools will be similar to that of tools that assemble individual scannable flip-flops or latches into serial paths within an IC. Where scan design is used at the IC level, it is common for the designer to focus solely on the task of interconnecting flip-flops, latches, and other logic such that the functional specification is met. During this stage of design, the need for the flip-flops and latches to be assembled into scan paths is ignored. Scan path assembly occurs once the functional design is complete and may be performed automatically by EDA tools. Note that, provided that the design rules for scan design are met, the ordering of the individual flip-flops and latches on the scan paths is largely arbitrary. The ordering can be selected to ease implementation, test development, and test application.

The ordering of boundary-scan ICs on the board-level scan path(s) can be similarly chosen to minimize costs.

3.12.2. Case 2: Clusters of boundary-scan ICs

As the number of boundary-scan ICs available to the designer increases, boards will begin to contain connections that go directly between such ICs (Figure 3.19). These connections can be tested using the boundary-scan paths and hence there is no need for them to be accessible to bed-of-nails probes.

With care, some probes can also be eliminated on connections that flow from a boundary-scan IC to one without boundary-scan, or vice versa. Cases where this can be done include:

☐ Networks where the connection feeds from a boundary-scan IC's output pin to one or more input pins on ICs without boundary-scan.

☐ Networks where the connection feeds from an IC without boundary-scan to an input pin on a boundary-scan IC.

For more complex networks, a probe will probably still be required.

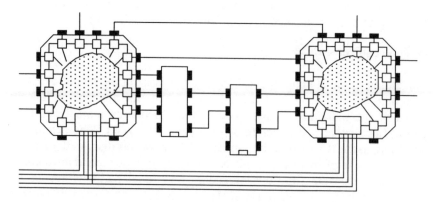

Figure 3.19 Clusters of boundary-scan ICs.

Test development and application requirements

Clearly, if probes are to be eliminated on boundary-scan-only connections, tools must be available that can generate tests for such connections. Also, the test system must be capable of applying these serial tests.

In cases where a probe is eliminated on a connection that includes an IC without boundary-scan, additional support tools will be required to ease the process of converting an in-circuit test module so that parts of it can be applied using boundary-scan paths (de Jong, 1990):

❏   parts of the test that would have been applied using the eliminated probe(s) must now be shifted into the appropriate boundary-scan cell(s);

❏   similarly, parts of the test result that would have been sensed using the eliminated probe(s) must now be observed by loading and shifting of appropriate boundary-scan cell(s).

Application of tests using the boundary-scan path must, of course, be synchronized to the conventional application of tests using bed-of-nails probes connected to signals that are not accessible via boundary-scan.

Design requirements

As more boundary-scan ICs become available on each board, the need for automated support for scan path assembly will increase, as outlined under Case 1.

Also, as the opportunity increases to eliminate bed-of-nails probes for some board interconnections, so the need for effective support tools for use during board layout rises. A tool is now needed that can identify where probes may be omitted and ensure that acceptable probe targets exist on all other interconnections. This tool will require a knowledge of:

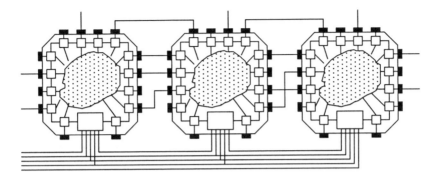

Figure 3.20   100% boundary-scan ICs.

❒   the circuit schematic;

❒   the PWB layout; and

❒   the boundary-scan characteristics of each boundary-scan IC (for example, as defined in its BSDL file).

3.12.3 Case 3: All ICs have boundary-scan

Once all the ICs on a board have boundary-scan, full advantage may be taken of the features defined by the standard — assuming that appropriate design and test development tools are available. These tools should include:

❒   a scan path assembler;

❒   boundary-scan test generation tools for chip-to-chip interconnect; and

❒   tools to assist in functional test of the loaded board, using boundary-scan to replace logic analysis and manual guided probing (Lefebvre, 1990).

Because boundary-scan can be used to test all chip-to-chip connections, there is no longer any need to ensure that the PWB layout can be probed. All restrictions placed on PWB layout to ensure suitability for in-circuit testing can be eliminated.

3.13. Conclusion

This chapter has provided an introduction to ANSI/IEEE Std 1149.1 and identified ways in which systems companies can reduce their test costs as boundary-scan ICs become more widely used. To be able to gain the full benefit from the boundary-scan circuitry that is available on a particular board design, tools are necessary to help in tasks such as PWB layout and test development. It is hoped that the EDA and automatic test industries will soon make such tools available.

References

Beenker F.P.M. (1985). Systematic and structured methods for digital board testing. In *Proc. IEEE International Test Conference,* Philadelphia PA, 380-5.

de Jong F. (1990). Boundary-scan test used at board level: moving towards reality. In *Proc. IEEE International Test Conference,* Washington DC, 235-42.

IEEE (1987). *VHDL Language Reference Manual (IEEE Std 1076).* New York: IEEE.

IEEE (1990). *Standard Test Access Port and Boundary-Scan Architecture (ANSI/IEEE Std 1149.1).* New York: IEEE.

Jarwala N. and Yau C.W. (1989). A unified theory for designing optimal test generation and diagnosis algorithms for board interconnects. In Proc. *IEEE International Test Conference,* Washington DC, 71-77. Reprinted in Maunder and Tulloss (1990).

Joint Electron Device Engineering Council (1986). *Standard Manufacturer's Identification Code.* (JEDEC Publication 106-A). Washington DC: JEDEC.

Konemann B. *et al.* (1979). Built-in logic block observation techniques. In *Proc. IEEE Test Conference Proceedings,* Philadelphia PA, 37-41.

Lefebvre M. (1990). Functional test and diagnosis: a proposed jtag sample mode tester. In *Proc. IEEE International Test Conference,* Washington DC, 294-303. Reprinted in Maunder and Tulloss (1990).

Maunder C.M. and Tulloss R.E. (1990). *An Introduction to the Boundary-Scan Standard.* Los Alamitos, CA: IEEE Computer Society Press.

Parker K. and Oresjo S. (1991). A language for describing boundary-scan devices. *Journal of Electronics Test: Theory and Applications (JETTA),* 2(1), 43-76.

Scholtz H.N. *et al.* (1988). ASIC implementations of boundary-scan and BIST. In *Proc. 8th International Custom Microelectronics Conference,* London, 43.1-43.9.

Yau C.W. and Jarwala N. (1989). A new framework for analyzing test generation and diagnosis algorithms. In *Proc. IEEE International Test Conference,* Washington DC, 63-70. Reprinted in Maunder and Tulloss (1990).

CHAPTER 4.
Planning for Design-for-Test

4.1. Introduction

Chapters 1 to 3 have discussed the need to design for testability and have described some of the most widely used design-for-test techniques.

This chapter considers how a development project can be managed so that the result is a testable design. As a preface to Part 2, the question of how to choose between the alternative design-for-test techniques is discussed.

4.2. Planning for a testable design

For the purposes of discussion, it is assumed that the development task for a new module or board design is composed of four sequential stages:

(1)   Product definition.

(2)   Architectural design.

(3)   Detailed design.

(4)   Transfer to manufacture.

The following sections indicate, from a test viewpoint, the tasks that should be performed in each stage.

4.2.1. Product definition

In this stage, the specification for the module or board is produced. This will detail the functions to be performed, the operating speed or throughput, and some aspects of its physical appearance — for example, the fact that a double-height EuroCard is to be used or the types of connector to be provided.

To allow initial planning for design-for-test, the following information should be included in the product specification:

☐   A definition of the function or functions to be performed, including the characteristics of and relationship between signals at its external interfaces.

☐   An estimate of the total number of units to be manufactured over the product life, with the quantity to be manufactured on a year-by-year basis if possible.

☐   Definitions of any features that must be included in the product to assist in the maintenance and diagnosis of the system of which it is a part. For example, to support field fault diagnosis to a replacable unit, it might be a requirement that a communications interface card (for example, Ethernet) be provided with various loop-back facilities.

☐   Definitions of any features to be included to allow the health of the module or board to be verified in the field (for example, power-up self-test).

☐   A statement of the assembly method to be used — surface-mount, dual-in-line/plated-through-hole.

☐   The expected performance of the production test program — the target fault types, the fault coverage, the run time (fault-free and including diagnosis), and the ATE types available.

☐   The 'budget' for design-for-testability. The aim here is to allocate a fraction of the total resources to design-for-test. For example, an amount of board area or a share of the total component cost should be allocated to design-for-test at the outset and relinquished for other use only if not required for that purpose. Too often, the designer uses all the available board space for circuitry required to meet the functional specification, with the result that none is left for design-for-test. Under these circumstances, little can be done to render the design testable. (The question of how much budget to allocate for design-for-testability is discussed further in Section 4.6.)

In addition, the target manufacturer and repair organization (if different) should be identified wherever possible. For high-volume products (say, 10,000+ units per annum) this is essential, because the greatest economy in design-for-test can be achieved only by tailoring the circuit design to the target ATE systems (see Section 4.5.3).

A checklist is included in the Appendix to help in recording the above information.

4.2.2. Architectural design

Alternative block-level designs for the circuit are explored and key design decisions are made. For example, a decision may be made on the microprocessor family to be used and/or whether custom ICs will be used. Increasingly, simulations are performed at a behavioural level as a part of this activity — for example, using VHDL (IEEE, 1987).

As a part of the design-for-test process, the following should be created and reviewed during this phase:

☐   An outline **bill of materials** (BOM) showing the key types of component that might be used (for example, the microprocessor family selected). This should be reviewed against any known testability requirements or known test problems, for example as advised by the target manufacturer based on prior experience. The Component Selection checklist in the Appendix can be used to assist in this review.

☐   Specifications for custom ICs, including a description of design features to be included to help test the loaded board — for example, ANSI/IEEE Std 1149.1 (IEEE, 1990).

☐   A test plan for the overall functional test or self-test of the complete product. This should detail the test to be performed, the way that the tests are to be applied, and the way that results will be observed. While it should not specify precise test stimuli or responses, it should show the routes to be followed to get this data through the circuit to or from each component or functional block. Consider, for example, the RAM block of a microprocessor-based board design. The part of the test plan that deals with this block might specify that the ATE will drive and sense data via the microprocessor bus and that the microprocessor and other components should be disabled during this stage of the test so that this can be achieved.

4.2.3. Detailed design

Detailed circuit schematics and board layouts are created, logic-level simulations are performed, and prototypes are constructed and debugged.

The testability of the design will be considered in detail during this stage and the following items should be created and reviewed:

❐   A final BOM showing the components used to construct the loaded board. As for the outline BOM, this should be reviewed against any known testability requirements or test problems.

❐   A complete documentation pack for the design, including a description of any design-for-test features added to help in testing. Why spend time and money on design-for-test if you're not going to tell the test engineer what you've done?

❐   Test waveforms for any custom ICs or programmable devices (for example, PLAs) included in the design.

❐   A functional test for the complete board design. This is the detailed implementation of the test plan created during the previous stage. The precise patterns of 1s and 0s that will be applied and sensed have now been computed.

4.2.4. Transfer to manufacture

The finished design is transferred to the selected manufacturer. The following test-related activities will occur during this stage:

❐   Identification of the target manufacturer and (where appropriate) repair organization, if not identified previously. Note that this decision can only be left until this late stage in cases where the design does not 'push' any limits. For example, it should not include any timing-critical signal paths, use novel components, or have smaller than average board geometries. It should, in fact, be a perfectly *average* design.

❐   Conversion of tests for custom ICs and programmable components and of the functional test for the complete board into the formats required by the target ATE.

❐   Extraction of data from the printed circuit layout to permit construction of a test fixture.

4.3. Testability checklists and design reviews

To help manage the process of designing a testable board, Part 2 of this book contains a detailed set of rules and guidelines. This material is supported by a set of checklists, included in the Appendix. These checklists should be completed by the designer during the appropriate stage (mostly, during detailed development) and used as an input to design reviews. The checklists allow the project manager or customer to verify that design-for-test issues have been properly considered and, where appropriate, that any violations have been signed off by competent staff.

4.3.1. 'Dear Designer ...'

It is appreciated that designers have many issues to consider while developing a new product and that they must often work within tight timescales and budget limits. For this reason, this book defines a design-for-test methodology that is as straightforward as possible. The aim is to ensure that designers can get on with the task in hand — the creation of a testable board design. For this reason, the various rules and guidelines have, as far as possible, been grouped on a task-by-task basis — for example, component selection, circuit design, and board layout.

This book is not an attempt to limit the designer's freedom to innovate.

Rather, its purpose is to help the designer develop a circuit that both meets the functional requirement and can be effectively manufactured, tested, and supported. Designs that meet the rules set out in Part 2 will usually be easier to debug. thus reducing the designer's task as well as that of the test engineer.

4.3.2. Rules and guidelines

Part 2 contains rules and guidelines that show how to design testable loaded boards. Each chapter focusses on one design activity — for example, component selection.

The rules given in each chapter must be followed wherever possible if a particular feature or component type is present in the design. They relate to issues that are critical to the ability to test the loaded board and should be violated only with the agreement of experienced test personnel.

In contrast, the guidelines provide freedom for the designer to make trade-offs between testability and other design criteria. They indicate problems that can occur when testing a loaded board and show how these problems can be avoided. Here, the aim is to give designers the information they need to make these trade-offs intelligently.

The reason for including each rule or guideline is given so that the designer can understand the potential impact of non-conformance.

4.3.3. Testability checklists

The rules and guidelines in each chapter are summarized in a checklist which will be found in the Appendix. The format of each of the checklists (with the exception of the management checklist) is similar to the example given in this section.

In each checklist the rules are listed first. *Rules must be followed if it is to be practical to test the board.* For each rule, the checklist allows the designer (or alternative completer) three options:

☐   *Yes.* The rule is complied with completely.

☐   *No.* The rule is violated in some cases. If this option is selected, an explanation of the non-compliance should be given on an attached sheet. The explanation can then be considered by the test engineer for the design. *The violation is acceptable only when agreed by the test engineer.*

☐   *N/A.* This rule is not applicable to the design — for example, because a circuit structure or component is not used.

Note that the rules are not rigid — the designer always has the option of convincing the test engineer that non-conformance is acceptable. The important point is that the guidelines and, more importantly, the rules should only be neglected where justifiable. An aim of the checklists is to ensure that design decisions that impact testability are properly considered and recorded.

Following the list of rules, guidelines are summarized. Because the designer can decide to make trade-offs between meeting the guidelines and achievement of other design objectives, the designer is allowed two options:

☐   *%.* The percentage of occasions on, or extent to, which the guideline has been followed. It is suggested that a return of less than 75% should be justified by an attached explanation as if this represented a rule violation.

☐   *N/A.* The guideline is not applicable.

Clearly, the greater the number of guidelines that is followed, the higher the testability of the finished design will be. It is therefore suggested that an average rating of 75% for all applicable guidelines should be requested before a design is considered suitable for manufacture.

4.3.4. Design reviews

A design review meeting should be held at least at the end of each development stage. This should be attended by the designer, the test engineer, and others to represent different interests in the design:

Component selection

(1) Product identity

Product		Version	

(2) Rules. The product *must* meet these requirements.

Item	Rule description	Ref.	Yes	No	N/A
1	All device-specific testability requirements have been implemented	7.2.3			

Note: a negative response must be justified on an attached sheet

(3) Guidelines. Meet these requirements where possible.

Item	Guideline description	Ref.	%	N/A
1	Components are in the approved components list for the target manufacturer	7.2.1		
2	Simulation models are available for component used	7.2.1		
3	ICT test data is available for components used	7.2.1		
4	Components used contribute to the 'buy testable' policy	7.2.2		

Note: a response of <75% should be justified on an attached sheet

(4) Sign-off

	Role	Signature	Name	Date
1	Designer			
2	Test engineer			
3	Project manager			

❒   Can it be manufactured?

❒   Does it meet safety requirements?

and should be chaired by the project manager or a neutral appointee.

The attendees should be given a period (one or two weeks) before the review meeting in which they can examine the design documentation — including the completed checklists. They may choose to meet with the designer during this period to discuss any points of concern.

At the review meeting, the designer should briefly review the structure of the design and the major design decisions made, with the objective of putting the subsequent discussion in context. The presentation should highlight any interesting or contentious issues — for example, the inclusion of a testability feature to assist in field fault diagnosis or the violation of a testability rule.

The reviewers should then be allowed to raise any matters of concern — particularly those whose solution may impact on other interests represented at the meeting.

The meeting should end with a record of the actions the designer should undertake to satisfy the reviewers' requirements. The expectation is that the designer will have met with the reviewers during the development process and discussed difficult issues with them. Therefore, actions agreed at the end of the review meeting should normally be minor.

The minutes of the review meeting should be added to the documentation for the design.

4.4. The test strategy

4.4.1. Test stages

In general, it will be necessary for a newly-manufactured board to undergo two stages of testing:

(1)   A *test for assembly-induced faults* to ensure that all components are in the correct locations and are correctly soldered to the board. Checks will also be made for other defects that may be introduced by the assembly process — for example, solder shorts between printed circuit tracks. This stage of testing can be performed using an in-circuit or cluster tester or via boundary-scan paths built into the ICs. Note that the ability of the complete board is not checked at this stage.

(2)   A *performance test* to verify that components interact correctly over the required range of clock speeds or frequencies. This stage of testing can be performed by plugging the board into a 'mock-up'

working system, by using a functional tester, or through self-test capability built into the board design.

A design should allow both 'assembly-oriented' and 'performance' testing to be performed at acceptable cost. The rules and guidelines in Part 2 are selected to ensure that this will be the case.

4.4.2. Structured or unstructured design-for-test

There is a widespread trend across the electronics industry towards use of structured built-in test techniques, such as boundary-scan and self-test, in place of tester-based techniques, such as in-circuit and functional test. The reasons for this were cited at the beginning of Chapter 2 — miniaturization is reducing test access and the operating speeds of loaded boards are escalating beyond the capability of the typical ATE systems. The rate of change to use of the structured techniques can be seen from Figure 4.1, which predicts the take-up of ANSI/IEEE Std 1149.1 (IEEE, 1990).

The design-for-test strategy presented in Part 2 follows this trend by advocating:

(1)   that ICs compatible with ANSI/IEEE Std 1149.1 are used wherever possible because this will reduce test development costs for all board designs and, where required, allow some internal connections to be accessed without using a bed-of-nails; and

(2)   that as much self-test capability is provided as possible, since this reduces the cost of 'performance' testing and provides a valuable health-check facility for use in the field (for example, during system-level fault diagnosis).

4.5. Choosing a design-for-test strategy

To give the most testable design at the lowest cost, the choice of which design-for-test technique or techniques to use for a particular board design must be determined by a number of factors. The most important factors, and the decisions made in developing the rules and guidelines in Part 2, are discussed in the following subsections.

4.5.1. The type of product

Designs will vary from low cost consumer products (such as telephones and calculators) to high cost 'capital' products (such as telephone exchanges and mainframe computers). They will include analogue and/or digital circuitry and operate at almost any frequency from D.C. to light.

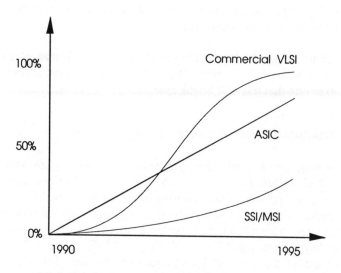

Figure 4.1   Take-up of ANSI/IEEE Std 1149.1.

Clearly, test techniques vary considerably across this spectrum of product types. Testability requirements will also vary considerably, in particular to 'match' the board design onto the available test equipment.

It is impossible to consider every possibility in this book. Therefore, the book concentrates on 'core' product types such as:

☐   100% digital board designs

☐   largely-digital designs where the analogue circuitry is located at the periphery (and is therefore readily separated from the digital circuitry).

If you have an 'unusual' product you will need to adapt the information given in this book to meet the particular requirements of the design and the test equipment. Ideally, you should involve a competent test engineer at an early stage.

4.5.2. The assembly technology

In Chapter 11, it is assumed that boards that do not use ICs compatible with ANSI/IEEE Std 1149.1 will be built using:

☐   through-hole components (for example, dual-in-line ICs) mounted on plated-through-hole boards; or

☐   surface-mount components spaced to give a low overall density of ICs on the board.

Where a high component packing density is required, ICs compatible with ANSI/IEEE Std 1149.1 will be used.

4.5.3. The production volume

In Chapter 1, it was mentioned that test costs fall into two types:

❑   Recurring costs that are incurred for each copy of the design manufactured — for example, the cost of components added to improve testability and the cost of applying the test.

❑   Non-recurring costs that are incurred once, usually during product development — for example, the cost of test generation.

For products manufactured in low and medium volumes (say, up to 10,000 units over the product's life), non-recurring costs dominate. In some cases, the sale price of the board will be determined exclusively by the development cost — the cost of the parts contributes a very small share of the total. In these cases, the development process (including the design-for-test approach) needs to be uniform and routine. One simply cannot afford to produce a custom test strategy for each design. Therefore, the aim is to have an approach that is flexible, allowing it to accommodate a wide range of design types, and be manufacturer independent.

The design-for-test rules and guidelines presented in Part 2 are designed with this aim in mind. Their goal is to keep test development costs and other non-recurring costs low, even if this results in slightly inefficient use of components or board real estate. In effect, the trade-off is between development time and unit cost — and this has been made in favour of reduced development time.

In contrast, for high-volume products (say, in excess of 100,000 units over the product's life), recurring costs dominate. The drive is therefore to minimize the parts count and the cost of components, even if this increases development costs. Significant savings can be gained by achieving an optimum match between the design, the manufacturing process, and the test equipment. To achieve these savings, the test engineer *must* be a member of the design team from the outset of the project. He or she must be intimately involved in the creation of the design and may wish to create a design-for-test strategy that is highly specific to the design. This will require that the manufacturer be identified at a very early stage, because the capabilities of the test equipment available will have a significant impact on the amount and nature of the design-for-test features that need to be added.

For high-volume products, the rules and guidelines in Part 2 are offered as a starting point for the design and test engineers. The explanations that accompany each rule or guideline will, it is hoped, allow the purpose of each to be understood and the problems that may arise from violation to be carefully assessed.

4.6. Setting a design-for-test budget

The amount that can be spent on design-for-test will, then, be determined primarily by the expected production volume.

4.6.1. Added circuitry

The majority of designs (perhaps 80% or more) are manufactured in low or medium volumes. As has already been mentioned, for such designs the principal objective is to reduce the non-recurring costs of development and test generation.

For integrated circuits, Toshiba have stated (Nozuyama *et al.*, 1988) that design-for-test features (such as scan paths and self-test) can account for up to 20% of the circuitry in a low volume ASIC. In contrast, for a high volume IC, such as a microprocessor, design-for-test circuitry would have to account for less than 5% of the total circuitry.

This principle applies equally to board designs — the share of the board's cost that is attributable to design-for-test can be higher for a low-volume product.

For a board design, much can be accomplished by careful selection of the components to be used to realize the intended function (see Chapter 7), so the number of components that will need to be added just to support testing can be kept low. This is especially true where components conform to ANSI/IEEE Std 1149.1, because these offer extensive test access to their input and output pins.

It is therefore suggested that a budget is set on the order of:

❏   *Additional ICs.* A number of ICs should be reserved for design-for-test purposes, as shown in Table 4.1. These added ICs will, in general, be relatively low cost devices. As will be discussed in Chapter 11, it is useful if a standard board layout 'template' can be used that provides fixed locations for the added ICs across the range of board designs that use the same board style (for example, EuroCard).

❏   *Additional resistors, etc.* Typically, a number of resistors and other discrete components will need to be added, for example to provide a pull-up to logic 1 that may be overridden by a tester. The added resistors may be grouped into a resistor pack when appropriate. Typically, on the order of 1 discrete components per IC is required for test purposes.

The addition of a connector to allow connection between the tester and test access points may also be justified.

Table 4.1 Number of ICs to be reserved for test purposes.

Number of ICs	Added ICs for test
1 to 25	0 or 1
26 to 50	1 or 2
51 to 100	2 or 3
over 100	3% to 5% of number

4.6.2. Added design time

Once the basic principles of design-for-test have been mastered and designers have become familiar with the various rules and guidelines, the additional design time needed to ensure a testable design will be low.

Indeed, the inclusion of design-for-test features may well reduce the overall design time for complex designs, because designers will find that the test access they have provided helps during debugging of prototypes.

References

IEEE (1987). *VHDL Language Reference Manual (IEEE Std 1076)*. New York: IEEE.

IEEE (1990). *Standard Test Access Port and Boundary-Scan Architecture (ANSI/IEEE Std 1149.1)*. New York: IEEE.

Nozuyama N., Nishimura A. and Iwamura J. (1988). Design for testability of a 32-bit microprocessor, the TX1. In *Proc. IEEE International Test Conference*, Washington DC, 172-182.

Part 2

Part 2 contains a set of rules and guidelines that will allow the development of a testable loaded board design.

The material is organized into chapters, each of which covers a particular design activity — for example, component selection or circuit design.

CHAPTER 5.
Test Access Techniques

5.1. Introduction

The subsequent chapters show how an adequately testable circuit design can be created. Some of the design practices described in these chapters specify that signals, normally internal to the circuit design, should be controllable and/or observable directly by the ATE system if adequate testability is to be achieved. This chapter provides a catalogue of commonly-used techniques for improving test access to printed circuit boards. The catalogue is by no means complete — many other techniques can be used, the objective being to provide test access at the lowest practicable cost.

Access to test connections to the product should be provided via one of the following (in order of preference):

(1)   The product's edge connector or other interface, if any 'spare' connections are available once the functional requirement is met.

(2)   Test access points accessible through a bed-of-nails test fixture (see Chapter 11 regarding physical placement of test access points).

(3)   Dedicated test connectors or sockets.

Figure 5.1 Test point symbol.

The requirement for physical test connections through one of the above can be reduced by adding components to the basic design, for example in situations where physical access is expensive (see Sections 5.4 and 5.6).

In this book, the symbol shown in Figure 5.1 is used to show a test connection to the circuit

5.2. Connector 'U' links

Where two or more spare pins are available on a connector, these can be used to separate the driving and receiving ends of a signal connection as shown in Figure 5.2. This will allow the connection to be controlled and observed during testing and, if necessary, for a different signal state to be applied to the input end compared to that generated within the circuit design.

During normal operation, a 'U' link on the backplane between the two connector pins completes the signal path. When the board is removed from the backplane (for example, during testing) the two halves of the signal are separated, allowing direct observation of the driven values and direct control of the receiving devices.

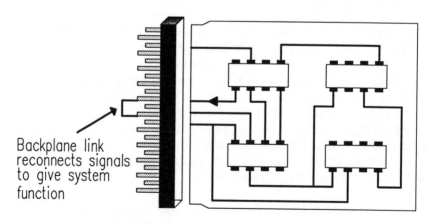

Backplane link
reconnects signals
to give system
function

Figure 5.2 Use of connector links.

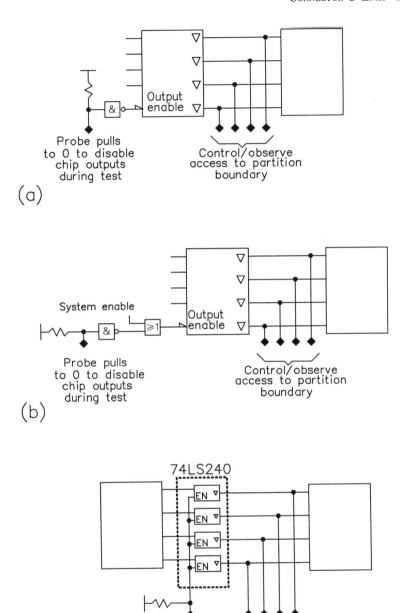

Figure 5.3 Use of 3-state outputs and buffers.

5.3. 3-state devices

If the circuit contains devices with 3-stateable outputs then this facility can be used to allow test access to the driven connections, in addition to any system requirement for the device outputs to be set to high-impedance (Figure 5.3a). The cost of achieving access in this way is typically very low, since many devices include 3-state outputs. If the 3-state capability is used to meet the system requirement, then often inclusion of an additional gate in the control line will make the facility usable during testing (Figure 5.3b).

Note that in bus-based designs (for example, microprocessor applications) the ability to control all devices connected to the bus so that the choice of bus driver is determined by the test system will allow the bus to be used for test access to all the blocks of logic to which it connects.

Where no suitable devices with 3-state outputs exist in the design, additional 3-state buffers (for example, 74LS240) can allow partitioning or test access (Figure 5.3c).

5.4. Multiplexors and shift-registers

For highly-miniaturized designs, where access using a bed-of-nails fixture is difficult, multiplexors and shift registers can be used to provide test access from the design's edge connector, etc.

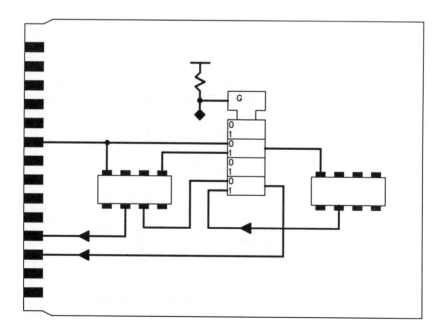

Figure 5.4 Use of multiplexors.

Multiplexors allow sections of logic to be by-passed during the appropriate phase of testing and for the sharing of connector pins for system and test access (Figure 5.4). They are particularly useful where large amounts of test data must be supplied or observed, or where the timing of test patterns is critical. Where a design contains a block of logic for which existing test data is available, the use of multiplexors to render its inputs and outputs accessible to the test system will permit the existing test data to be re-used without modification.

Shift-registers can provide a means of controlling or examining large numbers of points in a circuit through a small number of test connections. They are particularly useful for supplying test control signals, such as for enabling/disabling 3-state devices (see Section 5.3). Figure 5.5 shows two octal shift registers serially connected to provide access to:

☐ 8 signals in the circuit that control test operation, etc.

☐ 8 signals in the circuit that need to be observed during testing.

Note that the outputs of the serial-in, parallel-out shift register are fed to a parallel latch within the 74LS596. By clocking the output latch only when shifting is complete, the test control signals do not alter while a new pattern is being shifted in.

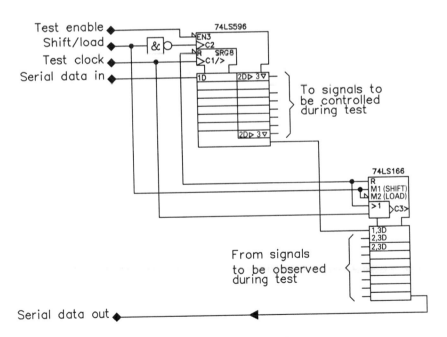

Figure 5.5 Use of shift registers

5.5. Use of simple logic gates

Simple logic gates can be used to inhibit signal flow as shown in Figure 5.6, for example to disable signal propagation around feedback loops. The gates can either be provided solely for test purposes, or advantage can be taken of gates already present in the signal paths (for example, by adding another input).

If, for example, a totem-pole TTL NAND gate is used (Figure 5.6a), then the output can be forced high by pulling the test access point low. This will permit the tester to inject signals at the gate's output using overdriving techniques (subject to time limits, etc.). *Note that this is not recommended,* and that the use of an open-collector gate (Figure 5.6b) is preferred since the output signal can be controlled without overdriving.

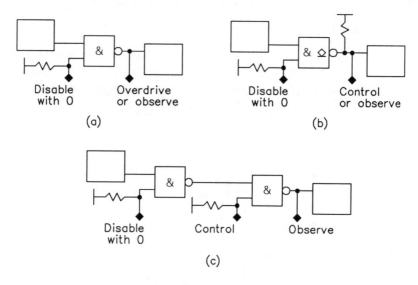

Disable   Overdrive
with 0   or observe

(a)

Disable   Control
with 0   or observe

(b)

Disable   Control   Observe
with 0

(c)

Figure 5.6   Use of simple logic gates.

Alternatively, potential overdriving problems can be avoided by using a pair of NAND gates in a multiplexor-like configuration (Figure 5.6c).

5.6. Test support chips

There are several commercially-available integrated circuits that are designed specifically to provide means of improving the testability of assembled printed circuit boards. For example, test support chips are available to:

- [ ] provide a means of controlling signals within the circuit design;
- [ ] provide a means of observing signls within the circuit design; and
- [ ] replace common building blocks (for example, octal flip-flops, octal bus buffers, and so on) with equivalent blocks that additionally provide test access.

Some examples of such components are:

- [ ] *AMD 29818*. This is an 8-bit register with supplementary test access facilities. The contents of the register can be swapped with the contents of an 8-bit shift register built into the component.
- [ ] *LSTI Testability Chip Set*. Members of this family of chips can be used to allow control or observation of internal board connections.
- [ ] *ANSI/IEEE Std 1149.1 Test-Support Chips*. Texas Instruments, National Semiconductor, and other companies offer 8-bit and 18-bit buffers, transceivers, latches, and flip-flops that conform to ANSI/IEEE Std 1149.1. Signals connected to these devices can be accessed through their boundary-scan paths using the instructions defined by the standard (see Chapter 3).

5.7. Bed-of-nails

The bed-of-nails fixture, which is used in in-circuit testing to apply tests directly to components in the design, can be used in functional testing as a low-cost means of observing connections internal to the board design.

CHAPTER 6.
Designing Self-Testing Products

6.1. Introduction

Many electronic products are intended for use in a domestic or office environment. They are of relatively low value (compared to, say, a mainframe computer or public telephone exchange). Many are produced in low to medium volumes — say, up to 10,000 units through the product's life.

Given this situation, there are clear advantages to the provision of self-test (or health-check) features within the products. For example:

❐   The equipment can perform a health-check when it is first turned on, and indicate any errors to the user.

❐   The user can run the health-check to help localize a fault. Information provided by the test can be passed to the repair organization, helping them to send the right technician with the right spares to fix the problem. In cases where a product is connected to others — for example, in a network — use of health-check tests can help in deciding which repair organization to call. A call to the wrong repair-man can prove expensive!

☐ The service technician can use the self-test along with portable instruments to help in locating the fault within the product.

☐ The self-test can be used as a key part of the post-production test. Here, the self-test can complement the in-circuit test approach used by many manufacturers. The self-test can be used as a product acceptance criterion.

That many companies are convinced of the value of these advantages is demonstrated by the number of small office systems which include self-test features: for example, computer terminals, personal computers, workstations, and printers.

Unfortunately, the means of designing a self-test capability into a product is somewhat of a black art, with the design being highly product specific. Therefore this chapter does not provide detailed rules for the design of self-testing products. Instead it provides a set of guidelines which will help in the design of self-test facilities.

6.2. Start small

Preferably, the self-test should be designed such that, in the event of a failure being detected, information can be provided on the possible location of the fault. For a multi-board product, this will provide the technician with a good starting point for repair — for example by replacing the board indicated as being faulty. For a single board design, the value is primarily in the repair facility rather than to the field technician.

The need to provide a degree of diagnosis dictates that a 'start small' strategy should be used for the self-test. That is, the test should initially check that a small kernel of components within the design is operational. In the next step this kernel can be used to apply a test to a limited amount of additional circuitry. If this circuitry also passes the test, it can then be used along with the kernel to test another block of circuitry, and so on until the complete design has been tested. This 'start small' approach is illustrated in Figure 6.1.

6.2.1. Microprocessor-based designs

For a microprocessor-based design, the kernel will typically contain the processor itself, the on-board clock generator, a small ROM containing

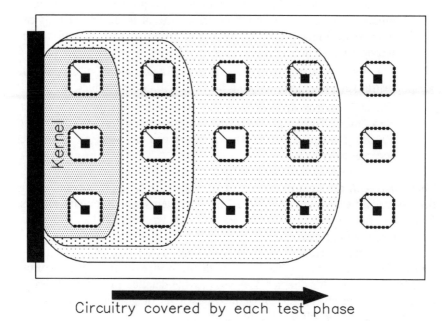

Circuitry covered by each test phase

Figure 6.1 The 'start small' strategy.

the self-test program, and associated components. Preferably, *a dedicated ROM should be used to store the self-test program* so that a clear distinction can be drawn between the kernel circuitry (which will perform the role of a tester) and the components to be tested (which will include the program memory).

Note that in microprocessor-based designs it is typical for many components to be connected onto the processor's bus, and this may give a problem in isolating the kernel for the first step in the test — checking the health of the kernel itself. In order to avoid this problem, it is recommended that the circuit is designed to allow the kernel circuitry to be electrically isolated from signals generated elsewhere in the product while the kernel test is executed. For example, in the bus environment, a bidirectional bus buffer can be included to isolate the segment in the kernel from the remainder of the bus.

The test of the kernel should include a health-check of the microprocessor itself which (as a minimum) verifies operation of each major internal register. For example, this could be achieved by writing checkerboard (1010...) and similar patterns into each register.

Following the kernel test, it is recommended that tests are performed on any on-board ROM or RAM (for example, as discussed in Section 6.6). By testing the memory early in the test procedure, it can then be made available for use as working space during subsequent tests.

6.3. Triggering self-test

The self-test should execute in response to events such as the following:

☐   power-up;

☐   a transition on a pin provided on the principal interface into the product (for example, its backplane connector) — for digital designs where 5 volt logic is used, the self-test should run following a change to 0 at the pin and an internal pull-up resistor should be provided to ensure that the pin is inactive when unterminated;

☐   a command received via a system interface (for example, via a network protocol or from the keyboard); or

☐   a push-button switch on the product.

A continuous or 'stop-on-fail' self-test mode can also be provided for use during burn-in of the product or when troubleshooting for intermittent faults.

6.4. Pass/fail indications

The results of the test should be indicated through one or more of the following :

☐   *An indicator light.* To ensure that the light is not itself faulty, the circuit should be designed (say) such that the light is on for a brief period at the start of the self-test (for example, during the kernel test) and will otherwise stay off unless a fault is detected. The 'brakes faulty' indicators on many cars are an example of this type of operation.

☐   *One or more connections at the product's principal interface* (for example, the backplane connector). If possible, this interface should both distinguish between pass and fail and provide some diagnostic information in the event of failure. The design must ensure that a fault on the 'status' connector cannot give the fault-free signal.

☐   *A VDU screen or other alphanumeric display built into the product.* In this case, the design should be such that a sequence of messages of the form:

> Starting KERNEL test
> KERNEL test passed
> Starting RAM test
> etc.

is displayed as the test proceeds, together with any diagnostic information which can be given on failure.

In all cases, where a fault is detected during the self-test, the failure indication given at the end of the test should be held until the test is re-run or some other command is given to restart the product's operation.

6.5. Control of external interfaces during self-test

While a self-test is executing within a unit (for example, a board, a set of boards, or a complete product) it is necessary to ensure that no erroneous data is supplied through its external interfaces. That is, the interfaces must be controlled such that it appears that the product is inactive.

It is particularly important that the circuit is designed such that no hazardous signal can be supplied to a connected product which is not involved in the self-test. (The term 'hazardous' is used here to cover situations which could risk the safety of equipment users and situations where connected products may be physically damaged.)

6.6. Component-specific self-test requirements

This section defines the types of tests which should be performed on certain specific circuit blocks.

6.6.1. ROM

The contents of the ROM built into a product may change at intervals as the firmware used is upgraded. In order to minimize the effects of such changes, the ROM should be designed such that a constant test result can be obtained regardless of the precise content. Figure 6.2 gives an example of how this can be achieved.

Assume that there are 2^N words of memory within the design, in which case the first $2^N - 1$ words will be available for program storage, and so on. while the last word is reserved. The test involves the generation of a checksum (for example, using cyclic redundancy coding techniques) for the data held in the first $2^N - 1$ words of memory. The resulting checksum is then compared with the data stored in the final word — which should be programmed to hold the expected fault-free checksum result.

6.6.2. RAM

A test of the RAM built into a product should be capable of detecting at least the following faults:

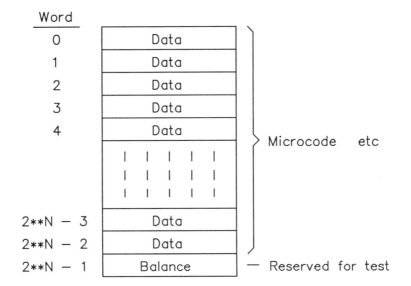

Figure 6.2 ROM data contents.

❒ any single memory bit being unable to hold either 0 or 1
❒ incorrect operation of the address decoder
❒ shorts between adjacent data input or output pins

An example of a suitable test procedure follows:

(1) Write data words into memory such that :

    (i) the word stored at address N is different from that stored in locations which have addresses one bit different from N (for example, the data in word 4 must be different from that in words 0, 6, and 5);

    (ii) the word stored at address N is different from those in addresses $N - 1$ and $N + 1$; and

    (iii) no word contains 111...1 or 000...0.

A suitable pattern could easily be generated using a linear-feedback shift register (or firmware analogue) and writing successive parallel outputs of the register into the memory as test data words.

(2) Form a checksum by reading the data stored in each memory word.

(3) Write the inverse of the previous data pattern into each memory word.

(4) Continue the checksumming process while the new data stored in each memory word is read.

(5)   Compare the final checksum with the expected result, which can be stored in the self-test program ROM.

The Joint Electron Device Engineering Council (JEDEC) are developing an industry standard for self-test features built into memory components. Components that offer this facility should be used in preference to others.

6.7. Some useful techniques

6.7.1.On-board test generation and signature analysis

Linear feedback shift registers (LFSRs) and multiple-input signature registers (MISRs) (see Chapter 2) can be built into a circuit, for example to provide a hardware mechanism for generating the data needed for a RAM test and creating the checksum from the results. Circuitry such as this is especially useful if placed on the principal buses in the design, since it can then be used to test many of the major component blocks.

Some of the ICs listed in Section 5.6 contain LFSRs and MISRs.

6.7.2.Loop-back

For products which have interfaces onto communications networks (for example, Ethernet, FDDI, and ISDN) a loop-back mode should be provided for use during self-test execution in which the 'send' data is fed immediately back into the 'receive' port.

The loop-back facility can be either internal or external to the product. In the latter case, the loop would be completed by an external cable connection between the two ports while in the former case additional hardware is built into the product to make the connection internally in response to a given control signal. *Internal loop-back is preferred.*

While the loop-back mode is selected, care must be taken to ensure that no invalid signals are applied to the network connected to the data-out port (see Section 6.5).

CHAPTER 7.

Component Selection and Design

7.1. Introduction

This chapter addresses factors which should be taken into account while selecting types of component to be used in a circuit design and while specifying and designing components.

It should be noted that the chapter focusses exclusively on the test aspects of component selection and that there are other factors which designers should also take into consideration when selecting components — for example, their suitability for automated assembly, and so on.

It should also be noted that the material in Section 7.4 is intended to be sufficient to help in specifying a testable application-specific IC (ASIC) design, not to be adequate for use during the ASIC design task.

7.2. Component selection

7.2.1. Range of device types

The range of component types used in a design, or across a range of designs, should be limited for many reasons, for example:

❏   to limit the number of component types stocked by a factory of repair depot; and

❏   to avoid the need to train support staff in the use of an excessive range of devices.

To the above must be added the need to limit test development costs (low for components used in previous products, high for new component types) and to ensure that simulation models are available to support test development work.

Where possible, components that are already on the **Approved Components List** (ACL) for the target manufacturer should be selected in preference to new component types. These components will probably have pre-existing test data and simulation models, so these will not have to be developed as a part of the current design project.

For components not on the target manufacturer's ACL, it is advisable to check with test engineering personnel before selecting them for inclusion in the design.

7.2.2.The 'buy testable' policy

Some components contain features which either make the testing of the components themselves easier or enhance the testability of the products in which they are used.

Examples of features which make components themselves highly-testable include:

❏   scan design; and

❏   self-test modes of operation (for example, a JEDEC-standard memory self-test facility).

These design-for-test techniques were discussed in Chapter 2.

Examples of features that enhance the testability of higher level products include:

❏   the ability to place all outputs in a high impedance state under the control of one input pin;

❏   serial shadow-register paths (for example, AMD/MMI 'Serial Shadow Register' product ranges); and

❏   standard test access port and boundary-scan register features designed to ANSI/IEEE Std 1149.1.

Where possible, devices which offer test-support should be used in preference to others.

Note: The SSI/MSI logic families contain variants of many functions. Care in selecting between these variants will help improve testability. For example, use a device with 3-state outputs in preference to one without, since the 3-state capability can be used to improve test access. (Figure 7.1).

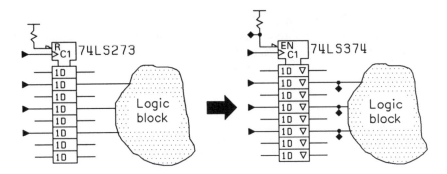

Figure 7.1   Choice of devices to improve testability.

7.2.3. Known test problems

Certain components may have specific test problems which are already known to the target manufacturer. Designers should check for the existence of known test problems for each component before including it in a product design and, *where restrictions in the use of a component are specified, these must be followed.*

7.3. Programmable device design

The following rules and guidelines should be followed when developing a programmable device.

7.3.1. Initialization

It must be possible to set the device into a known state by application of a simple waveform at one or more inputs.

Ideally, the device should be provided with a single asynchronous or synchronous reset input which, when the correct signal is applied, causes every register or latch, and so on, to be set to a known state.

If this is not achievable, then initialization must be achieved by an alternative means within the constraints specified in Section 8.2.

7.3.2. Ability to 3-state output pins

The device should be provided with an input which, when activated, causes all output pins to be set to a high impedance (inactive drive) state. Where this capability is not needed to meet functional requirements, then a dedicated test pin and associated circuitry should be provided. (See Figure 7.2.)

This facility eliminates the possibility of damage to the device from overdriving during in-circuit testing (see Chapter 1). A number of commercial ICs include this facility, controlled from dedicated test pins, for this reason (for example, the Texas Instruments TMS380 chip set).

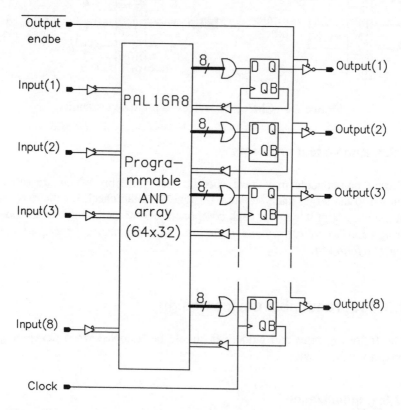

Figure 7.2 Achieving high-impedance outputs on programmable logic.

Note also that some programmable logic devices can impede in-circuit testing unless their outputs can be set into a high impedance state. The key problem is that the output drivers can become unstable when over-driving techniques are used to inject signals from the ATE. The resulting oscillations at the over-driven connection prevent the test being applied.

7.3.3. Avoid asynchronous designs

Asynchronous stored-state circuits (for example, asynchronous finite state machines, etc.) cause significant test and reliability problems and are best avoided completely. *Use a synchronous design instead.* If this is not possible, restrict asynchronous circuitry to a small part of the design which can be isolated from the remainder during test (for example, using techniques such as those catalogued in Chapter 5).

A key test problem is caused by the rapid propagation of signals around asynchronous feedback loops. For faulty circuits, this capability allows the effects of a fault to appear instantaneously at all points in the loop, making diagnosis of the cause of the fault extremely difficult.

Reliability problems can arise due to critical dependence on the timing properties of the components in the feedback loop. For example, if combinational logic is used to control an asynchronous reset of a stored-state device, then it can cause spikes, and so on, which cause unwanted clearing unless it is carefully designed. Reliability problems of this sort not only impact the performance of the circuit in use, they cause mis-operation during testing with consequent 'fault not found' during diagnosis.

Note: An exception to the rule is the use of asynchronous preset/clear/etc. inputs for circuit initialization (see Section 7.3.1).

7.4. ASIC design

The purpose of this section is to help the board designer to specify custom ICs in an informed manner, so that they include features which will secure board testability. Failure to follow the requirements set out in this chapter could significantly reduce the testability of the finished board design.

7.4.1. Initialization

It must be possible to set the IC into a known state by application of a simple waveform at one or more inputs.

Ideally, the IC should be provided with a single asynchronous or synchronous reset input which, when the correct signal is applied, causes every part of the design to be set to a known state.

If this is not achievable, then initialization must be achieved by an alternative means within the constraints specified in Section 8.2.

7.4.2. ASIC features needed to help in board test

Ability to force an inactive drive state at output pins

The IC must be designed such that all output pins can be set to a high-impedance or inactive drive state when an appropriate condition (input signal, instruction, and so on) is applied. Where this capability is not needed to meet functional requirements, then it must be provided solely for test purposes. It is essential that output pins can be placed in the high-impedance (inactive drive) state while in-circuit testing of adjacent components is in progress (see Chapter 1). Control of the facility can be achieved either through a dedicated test pin or through a specific instruction applied to the chip via an ANSI/IEEE Std 1149.1 Test Access Port (see Chapter 3).

ANSI/IEEE Std 1149.1

All ASIC designs should include design-for-test features in accordance with ANSI/IEEE Std 1149.1, Test Access Port and Boundary-Scan Architecture.

7.4.3. Test quality targets

A high quality test programme must be available for each custom IC.

Why is a high quality test needed?

Like printed circuit boards, custom integrated circuits must be tested to a sufficiently high standard following production to ensure an acceptable shipped quality level. The result of an inadequate test is that faults may remain undetected in components shipped for assembly onto a printed circuit board. Such 'dormant' faults may be detected either when the assembled board is tested or, in the worse case, may cause intermittent failure of a system in the field. Many companies have estimated that it costs 1000 times as much to locate and replace a faulty IC in the field as it would have cost to find the defect immediately following chip production.

What faults should the test detect?

The following fault types must be included in the target fault set:

(1) *Stuck-at faults on device outputs.* These faults represent device outputs becoming fixed at 0 or 1. Note also that the possibility of device outputs becoming stuck-at-Z (high impedance) should be

considered for devices with 3-state or bidirectional pins, but may lead to 'potential' detection (see below) unless the design ensures that the bus is pulled to 0 or 1 when it is not driven from another source. These faults all apply to a complete connection between devices — the fault is seen by all devices fed from the connection.

(2)   *Stuck-at faults on device inputs.* These faults represent individual device inputs becoming stuck at 0 or 1. The faults only affect the specific input and not other devices driven from the connection. They model defects in the fan-out branches of an interconnection — for example, an open-circuit in the segment of track feeding one gate input. In some simulators these faults are simulated as stuck-at faults, while in others they are simulated as open-circuit faults where the disconnected side of the connection is coerced to 0 or 1.

Note that a 'device' may be a logic gate or more complex cell in a semi-custom IC, or a transistor in a full-custom IC, depending on the level at which the circuit is modelled for simulation.

Optionally, the following class of faults may also be considered if time and budget permit:

(3)   *Bridging faults between adjacent device pins.* These faults can be used to simulate solder shorts, etc. between adjacent terminals of a device or between adjacent tracks.

How many of the target faults should the test detect?

The target is for the test programme to allow *all* faults in the target fault list to be assigned into one of three categories, as discussed below. *Under no circumstances should the number of faults which cannot be categorized exceed 5% of those in the target fault list.*

A fault is deemed to have been detected by a test programme as a result of one of the following:

(1)   *'Hard' detection.* A hard detection occurs when a fault causes a change from 0 to 1 (or vice-versa) at one or more points monitored by the external test equipment.

(2)   *'Potential' detection.* Potential detection occurs when a fault causes a predefined number of changes from either 0 to X (unknown) or from 1 to X at one or more points monitored by the external test equipment. The number of observed changes before detection can be adjusted to change the confidence in the detection of the fault, since the observed unknown signal state could be the fault-free value or its complement. A figure of 5 observed changes to X is typical.

(3)   *Acceptable non-detection.* There will typically be some faults for which detection is not possible due to the nature of the circuit design. For example, where a circuit contains redundancy, not all faults will

be detectable since correct operation of one part of the circuit will prevent the incorrect operation of another from causing failure at a point observable to the ATE. A further example would be a stuck-at-1 fault on a device input which is tied to the logic 1 since it is not required in the particular design. 'Acceptable non-detection' includes all cases where there is a clear reason why hard or potential detection is not possible.

A pitfall you should avoid

Normally, test quality will be verified using a fault simulator. However another technique — **'node toggling'** — is sometimes advocated by silicon vendors or design houses. In the node toggling technique, a check is made on a simulation of a test programme to ensure that each connection (node) is at some point set to both 0 and 1. Clearly, if a node does not get set (say) to 0 then the test programme cannot detect stuck-at-1 faults. However, simply setting the node to both 0 and 1 is not sufficient to ensure that faults on it are detected — the effect of the fault must also be made visible to the ATE at the component's outputs. Detection of the fault can *only* be guaranteed using fault simulation.

7.4.4. Internal testability

As for printed circuit boards, test costs for custom integrated circuits can be a significant part — sometimes as much as 50% even for relatively testable designs — of the total development cost, particularly for complex full-custom ICs. Clearly, costs of this magnitude can have a significant impact on the viability of using custom silicon in a product.

For this reason, *it is recommended that highly-structured design-for-test techniques such as scan design and self-test are used wherever possible* (see Chapter 2). These techniques can, in some cases, allow the test development task to be fully automated, thereby producing significant manpower savings. However, costs are incurred due to the need to dedicate a small number of package pins to test functions and due to increases in the physical size of the IC caused by test circuitry added to the design.

7.4.5. A typical testability budget for IC design

The amount which can be spent on design-for-test in an integrated circuit will depend on the quantity of devices which will be made over the production life. For high volume parts (such as microprocessors), test development costs per device are lower than for low volume parts (such as many semi-custom ASICs), so the amount which it is economical to spend on design-for-test is also lower.

For a low volume part (say, up to 2000 for total production quantity), a typical budget would be:

☐   Additional circuitry — up to 15% increase in the number of gates.

☐   Additional pins — 5, to allow provision of ANSI/IEEE Std 1149.1. (Note: Most test functions can be controlled through the interface defined by this standard.)

☐   Reduced performance — up to 5% of maximum operating speed.

For a high-volume part, the budgets for additional circuitry might be reduced to 5%. (Toshiba quote a variation between 20% for low-volume parts and 5% for high-volume parts within their company.)

Beware: if you do not allow an appropriate budget for testability (for example, the entire gate capacity is needed to achieve the system function) then you may eventually get an adequately tested component, but the cost will be high.

CHAPTER 8.
Circuit Design

8.1. Introduction

This chapter provides design requirements which will help achieve testability in digital circuit designs, or in the digital portions of mixed analogue/digital circuit designs.

The goals of the requirements in this chapter are to ensure that:

❐   design practices known to lead to testability problems are avoided; and

❐   the controllability and observability of internal signals are adequate by providing test access at key internal connections.

Adherence to the standards defined in this chapter will:

☐ reduce test development costs;

☐ reduce operator interaction during production or maintenance test, thus reducing test run times and increasing tester capacity; and

☐ ensure synchronization between the circuit and the test system.

8.2. Initialization

It is essential that the circuit can be set to a fully defined start state before tests are applied — that is it must be able to be initialized. If this cannot be done then the results of the test may not be repeatable, leading to low confidence in the test results or fault diagnosis.

This section discusses the need for initialization and the problems which arise if it cannot be achieved effectively. It also defines recommended techniques achieving initialization.

8.2.1. Why initialize?

The state of the circuit is defined by the logic values held in each stored-state device — flip-flop, counter, memory, and so on. Complete initialization is achieved when a known pattern of 0s or 1s has been written into each such device. This must be achieved as quickly and as simply as possible for two key reasons.

Firstly, some of the faults which can occur in a design will prevent it from initializing. For example, if there is an open circuit fault in the connection to a reset input on a flip-flop then applying the reset condition will have no effect. If the circuit can be initialized by applying a small number of input patterns, then the number of such faults will be relatively small. However, if the initialization process is complex, then the number of faults which prevent completion will be much greater. While this does not impact the test's capability to detect faults, faults which inhibit initialization are extremely difficult to diagnose, thereby impeding repair.

Secondly, initialization is vital during both design verification and the test development process. Here it is the cost of simulation which is adversely affected if initialization cannot be achieved, or is more complex than necessary.

Both fault-free simulation (design verification) and (to a greater extent) fault simulation run times will be increased if initialization is inefficient.

8.2.2. Length of initialization waveform

To limit the number of faults which can prevent successful initialization (and which are in consequence difficult to diagnose), the length of the initialization waveform (as measured by the number of clocks or input patterns applied) must be kept to a minimum. As a guide, complete initialization should be achieved within 20 clocks or input patterns.

8.2.3. Preferred initialization techniques

The following sub-sections discuss preferred techniques for achieving circuit initialization. They are organized to indicate relative priority between the techniques, with the first being the ideal option.

Asynchronous initialization

The ideal way of initializing a circuit is to apply a single pattern at the design's inputs which asynchronously initializes all stored-state devices. See, for example, Figure 8.1 where a reset signal is applied to the asynchronous clear inputs of all flip-flops in the design.

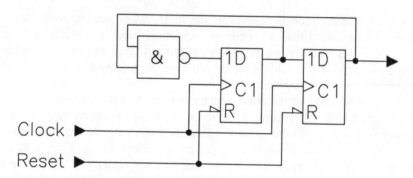

Figure 8.1 Circuit with asynchronous reset.

Synchronous initialization

A circuit can be synchronously initialized by applying a pattern of control inputs followed by one or more clock pulses. See, for example, Figure 8.2.

Initialization through shifting or loading of data

For highly regular designs such as parallel latches, long shift registers, and random access memory, asynchronous or synchronous initialization may be impractical. The provision of a clear or other input on each stored-state device is not practical due to the significant increase in circuit size (at the

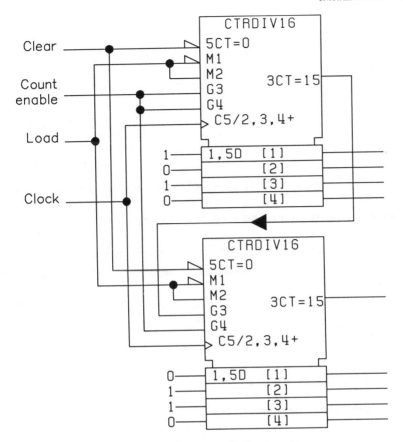

Figure 8.2 Circuit with synchronous reset.

chip level) which would be required. In these cases initialization must be achieved by shifting or loading data into the design.

8.2.4. Prohibited initialization techniques

The following sub-sections define initialization techniques which are not acceptable, for example due to the increased complexity of test development or use.

Homing sequences (Repeat-until)

Figure 8.3 shows a circuit which can be set to a known starting state by continuing to apply a clock until the required pattern appears at the circuit outputs. When this pattern is detected the test branches out of the initializing loop.

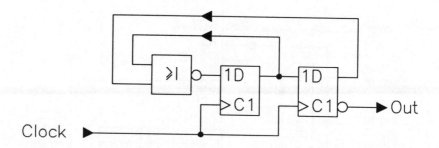

Figure 8.3 A circuit which requires a homing sequence for initialization.

This style of initialization is not acceptable because:

☐ most logic simulators do not have test-and-branch capability;

☐ the ATE cannot be guaranteed to have test-and-branch capability; and

☐ faults can easily prevent initialization and can cause indefinite looping unless a limit is set on the number of clocks, etc. to be applied.

8.2.5. Controlling initialization

Connection to initialization controls

The design must allow initialization to be controlled by application of defined signals to one or more inputs on a normal functional interface — for example, the edge connector.

Dedicated test connections to the design (for example, bed-of-nails access through test lands) may be used to enhance the initialization capability provided through the normal functional interfaces, but cannot be used as the sole means of initializing the circuit.

Power-on resets

Where power-on resets are included to meet a design requirement, the circuit must provide an alternative input through which the ATE may cause initialization without having to disconnect power. Figure 8.4 shows the provision of a dedicated test input to complement initialization at power-up.

If a power-on reset is used then time must be allowed for the output of the power supply to decay before it can be reapplied in order for the reset to be reliable. The requirement for an alternative means of triggering the reset allows test and diagnosis time to be reduced where it is necessary to initialize the circuit at several points during the test — not just at the

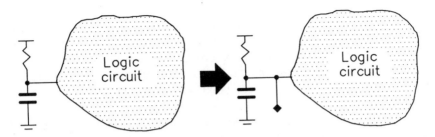

Figure 8.4 Power-on reset with test input.

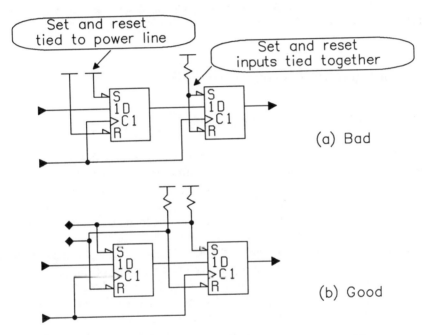

Figure 8.5 Methods of tying unused initialization control inputs.

beginning — for example, to allow the test to be structured. Also, during guided probing for fault diagnosis the test will need to be repeated frequently — a process which would be considerably slower without the capability to cause a 'logical' reset.

Unused initialization controls

Unused initialization-control inputs (for example, load, preset, clear, and so on) must not be tied directly to power or ground. In addition, where there are several initialization controls for one component then these must not be connected together (see also Section 8.5.4).

Figure 8.5 shows methods of tying such inputs in TTL-compatible circuits which satisfy the first requirement. Connections to be tied high for normal operation should be made through a suitable pull-up resistor. Connections to be tied low should be driven from the output of a gate (for example, an inverter) whose input is pulled high. In either case, a dedicated test connection to the test access point will allow the component to be initialized.

The requirement that unused initialization controls must not be tied together is included to ensure determinate behaviour when the initialization condition is removed. For example, if the preset and clear inputs to a flip-flop are connected together then the state after the combined signal changes from 0 to 1 will vary depending on propagation delays, and so on, in the circuit (Figure 8.5a).

8.3. Architectural issues

8.3.1. Mixed analogue/digital circuits

Different test techniques are required for analogue and digital circuitry. While testing of each type of circuitry in isolation is relatively straightforward, significant test problems can arise when attempting to test mixed analogue/digital designs.

An an example, consider the case of an analogue-to-digital (A-to-D) converter feeding into a complex digital circuit. Because of the performance tolerances inherent in the design of the A-to-D converter, a defined voltage applied at the analogue input may produce one of a range of digital patterns at the converter's output. If the converter's outputs are not directly accessible to the test system, then they will need to be propagated through the digital circuitry to signals which are connected to the ATE. This task is not necessarily straightforward, particularly if the individual output bits are processed separately.

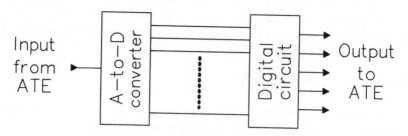

Figure 8.6   Testing a mixed analogue/digital circuit.

The variability of the A-to-D conversion process also impacts the task of testing the digital circuit. Digital circuit testing requires that defined inputs are applied, and that corresponding predefined outputs are observed. If the characteristics of the A-to-D converter cannot be accurately defined, then these requirements can only be achieved if the test system can control the inputs to the digital circuit directly, bypassing the converter.

Test access must therefore be provided as close to the analogue/digital interface as possible — for example as shown in Figure 8.7.

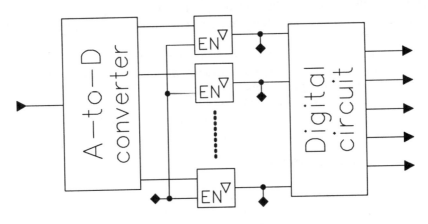

Figure 8.7 Partition analogue and digital circuits.

8.3.2. Dynamic circuits

Dynamic circuits must be operated at a minimum speed to avoid unintentional loss of data. There are, however, cases where this is difficult to achieve during testing — for example, when application of the test must pause to allow the ATE to fetch more data from backing memory.

Such problems may be overcome through use of keep-alive clocks supplied either from the ATE or by circuitry within the product design. However, since the length of time between bursts of test activity cannot be accurately defined, the circuit must be capable of entering a hold state where no data changes occur for keep-alive clocks to be useful.

Where possible, *use a static design* and avoid these problems.

8.3.3. Asynchronous circuits

Asynchronous stored-state circuits (for example, RS-latches, asynchronous finite state machines, and so on) cause significant test and reliability problems and are best avoided completely. *Use a synchronous design instead.* If this is not possible, restrict asynchronous circuitry to a small part of the design which can be isolated from the remainder during test (for example, using techniques such as those catalogued in Chapter 5).

A key test problem is caused by the rapid propagation of signals around asynchronous feedback loops. For faulty circuits, this capability allows the effects of a fault to appear instantaneously at all points in the loop, making diagnosis of the cause of the fault extremely difficult.

Reliability problems can arise due to critical dependence on the timing properties of the components in the feedback loop. For example if combinational logic is used to control an asynchronous reset of a stored-state device, then it can generate spikes that could cause unwanted clearing unless it is carefully designed. (The combinational network in Figure 8.9a includes a logically-redundant gate whose purpose is to prevent glitches being applied to the flip-flop input.)

Reliability problems of this sort not only impact the performance of the circuit in use, they cause mis-operation during testing with consequent 'fault not found' during diagnosis.

Note: An exception to the rule is the use of asynchronous preset, clear, and other inputs for circuit initialization (see Section 8.2).

8.3.4. Redundant circuits

Logically redundant circuitry may be included in a design for several purposes. Key examples are reliability improvement and removal of hazards in combinational logic networks. In other cases, redundant circuitry may be included accidentally — for example, by use of a four-bit counter stage which can only count up to 7 because of constraints imposed by the surrounding circuitry.

Do not build redundant circuitry into a design unless it really is necessary. Where it is needed, then the design must allow each redundant block to be thoroughly tested in isolation (see, for example, Figures 8.8 and 8.9). Failure to do this will, in general, act against the original purpose of including the redundancy — that is, it will make the overall result less reliable, or more susceptible to hazards, and so on.

In Figure 8.8, the voter in circuit (a) is intended to masks faults at the output of any one of the logic blocks at its inputs. It also does this during test, preventing detection of logic block faults at the circuit output. In circuit (b), test access points and 3-state buffers have been inserted between the logic blocks and the voter to allow:

❏   each logic block to be fully tested in isolation from the others; and

❏   the voter to be fully tested in isolation from the logic blocks.

In Figure 8.9, the top circuit contains a redundant term (highlighted gate output) to ensure that no glitches are fed to the asynchronous set input

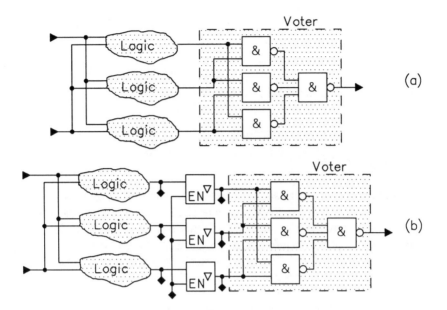

Figure 8.8 Fault-tolerant redundant design.

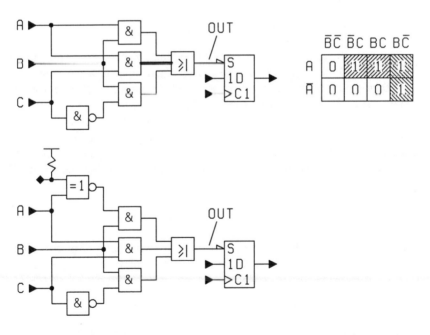

Figure 8.9 Redundancy for glitch suppression.

to the flip-flop. Without the redundant term, a glitch might be generated by the combinational network when moving between ABC=111 and ABC=110, for example.

8.4. Function-oriented requirements

8.4.1. Random-access memory

Access requirements

For thorough testing, test data must be written and read several times from each RAM word. Because RAM sizes are large, and increasing, the amount of test data required is also large. Any limitations in the ability to feed data in or read data out can therefore lead to significant test problems.

For example, a limited test for stuck-at faults in a 1K word RAM would require 3072 test patterns:

- ☐ 1024 patterns to write the first set of words (for example, a unique bit pattern in each word) and thus initialize the memory
- ☐ 1024 patterns to read back the first set of words while writing the second (inverse) set
- ☐ 1024 patterns to read the second set of words.

This assumes direct access to the RAM's terminals. If the surrounding circuit is such that 10 clocks are needed to deliver each input test and read each result then the number of test vectors required becomes 30,720. Clearly, for modern memories with >1 Mbit capacity, the inability to apply a test directly to the RAM data and address lines can present enormous problems. It is therefore a requirement that *all* RAM terminals (including the data and address lines, chip selects, and read/write controls) must be directly accessible to the tester, through test access facilities built into the design if appropriate.

Dynamic devices

The use of static RAMs is preferred. Where dynamic RAMs are used, internal refresh signals must be controllable from test points during testing.

8.4.2. Read-only memory

Testing the ROM

ROM is tested by exhaustive examination of the stored data, either by

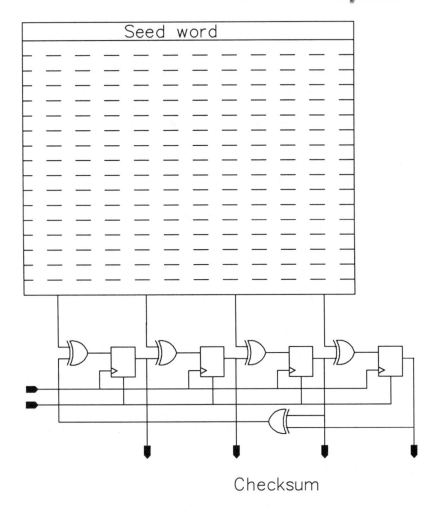

Checksum

Figure 8.10 Provide a seed word to ensure a constant checksum.

examining each word separately as it is read or by forming a checksum. In either case, it is necessary to be able to sequence through all memory addresses — for example, by applying a count to the address lines. If the basic circuit design does not allow this (perhaps because the ROM has been used as combinational logic — for example, to build a state machine), then test access must be provided to allow such a count to be applied.

The memory outputs must be accessible at test points either to allow examination of each word or to allow the checksum to be created.

Where ROM contents can change once the memory is assembled into the product (for example, where selection of a ROM is used to program the circuit function) provision must be made to ensure that all variants of the design can be tested with substantially the same test program. For example:

(1)   If the ROM is electrically reprogrammable, allow the contents to be erased and rewritten *in situ* without damaging surrounding components.

(2)   Provide a 'seed' word that ensures that the checksum result is constant regardless of the memory contents. Figure 8.10 shows how the use of a seed word can convert the checksum for the rest of the memory contents into a required result. (Information on checksum generators is contained in several of the texts referenced at the end of Chapter 1.)

Testing the rest of the design

A common technique for testing microprocessor-based board designs is to bypass the on-board ROM during appropriate parts of the test program and instead provide a ROM emulation in the ATE. This allows the rest of the product to execute a program supplied by the ATE, different from the one it will execute normally.

There are two main techniques for achieving this:

❒   Include a small section of code in the ROM which will be executed following reset of the microprocessor. This code should cause an conditional jump to an unused address. The code can inspect a logic level applied at a control input to determine whether the product is under test.

❒   Modify the address decoding logic of the product so that the ROM components are permanently disabled when the test control is active.

8.4.3. Monostables

Monostables must not be used except in exceptional circumstances because they cause considerable test problems — for example due to variable output pulse widths and their sensitivity to transients on input signals.

Where monostables have to be used to meet a design requirement the following requirements must be satisfied (Figure 8.11):

(1)   Monostables must not be cascaded.

(2)   Inputs and outputs of monostables must be connected to test points to allow accurate measurement of output pulse widths, and so on.

(3)   Test access must be provided to allow the component's outputs to be replaced by inputs from test points.

(4)   In cases where the output pulse is short (less than 300 nanoseconds) provision must be made to allow easy detection of the output pulse, either by lengthening the pulse during test or by including a 'glitch-capture' circuit.

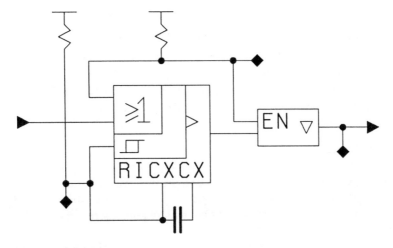

Figure 8.11 Testability requirements for monostables.

8.4.4. Clock generators

The following requirements ensure that the circuit can be properly synchronized to the ATE during testing and, in cases where the normal operating speed of the circuit is in excess of the ATE's, that testing can be performed at a speed acceptable to the test system (Figure 8.12). For in-circuit testing, the requirements also ensure that the tester does not need to over-drive a changing signal (which could cause unwanted spikes during the testing of devices supplied by the clock).

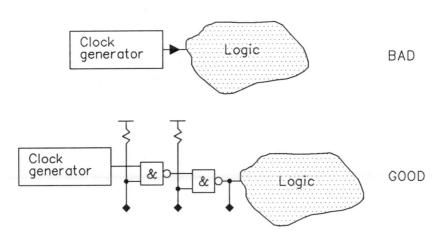

Figure 8.12 Clock generators.

(1)   Provision must be made for externally supplied signals to replace the outputs of built-in clock generators during test.

(2)   The outputs of built-in clock generators must be directly observable at a test point.

(3)   *The use of multiple free-running built-in clocks is prohibited* unless they can be independently disabled and replaced during testing.

8.4.5. Counters

Where long counters are included in designs they can result in an excessive number of test patterns being required for testing. This will have a direct impact on test costs, since the design will require more time on the test system and reduce throughput.

The example in Figure 8.13 shows a common use of a long counter — to reduce the frequency of a supplied clock before it is applied to other logic. In this case, the counter stages form a part of a digital watch chip.

In the top example, 86,400 clocks must be applied at the clock input to change the state of the programmable 'days-in-a-month' divider. This would lead to an excessive test time, even if the input clock frequency was increased from 1 Hz to 10 MHz during test (see Section 8.4.4).

To avoid such excessive test lengths, the design should allow long counters to be modified during test such that no more than ten stages precede any output from the counter chain, for example as shown in the modified circuit in Figure 8.13. In this way, the majority of the testing can be done

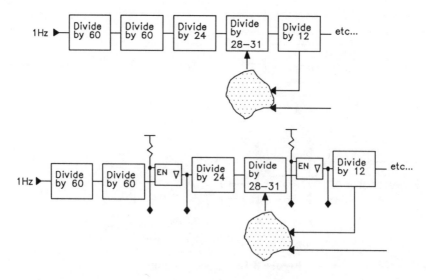

Figure 8.13 Long counters.

while bypassing the counter chain, with a small number of tests being applied with the complete counter chain connected to verify overall circuit operation.

For very long counters (say more than 16 stages) it is advisable to allow the chain to be divided into several shorter segments during testing so that a thorough test can be performed on the counter within an acceptable test time.

8.4.6. Parity trees

Parity trees should be constructed to allow segments of up to eight stages to be tested separately, and for outputs to other circuitry to be controlled directly from test points.

This will considerably simplify the task of changing the state at the output of the tree, which will be necessary when testing the logic this drives.

8.4.7. Adjust-on-test and select-on-test components

Adjust-on-test (AOT) and **select-on-test** (SOT) components (for example, as used to set the pulse width of a monostable) should be avoided where possible. Their use increases the amount of operator interaction required during testing and significantly lengthens test times.

Where AOT or SOT components must be used, then the following requirements must be satisfied:

(1)   No initial setting or selection should totally inhibit circuit operation.

(2)   The adjustment or selection required must be dictated by a rational electrical measurement made at a test point.

(3)   Interdependent AOT/SOT components are only permitted in exceptional circumstances where no other alternative is available. The number of interdependent AOT/SOT components must be kept to the absolute minimum.

(4)   A detailed adjustment or selection procedure must be supplied as a part of the design documentation.

(5)   For AOT components, the adjustment must be achievable without the use of unusual or specially designed tools.

Additional requirements on the positioning and mounting of AOT and SOT components on printed circuit boards are given in Chapter 11.

8.4.8. Switches

Where switches are used the time required for testing is increased since operator intervention is required to change settings.

Alternatives such as 'handbag' links (that is, as shown in Figure 8.14) that can be completely removed during testing are preferred since, with the link removed, the connections can be easily controlled and/or observed.

To allow testing to proceed at high speed where switches are used, access should be provided to allow the tester to control their outputs during test. This will allow the majority of testing to proceed without operator intervention, with a small number of checks involving the operator to verify the performance of the switches themselves.

See Chapter 11 for requirements on the orientation and mounting of switches.

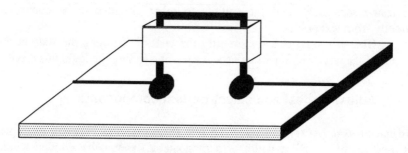

Figure 8.14 'Handbag' link.

8.4.9. Indicators and displays

Test points should be provided to allow signals feeding indicators (LEDs, and so on) and displays to be examined directly by the test system. This will allow testing to proceed at high speed without operator intervention, with a small number of checks involving the operator to verify the performance of the displays and indicators.

See Chapter 11 for requirements on the orientation and mounting of indicators and displays.

8.5. Connection-oriented requirements

8.5.1. Limit fan-out

The load placed on any component output by other components in the design must allow sufficient margin for additional loading presented by guided or bed-of-nails probes (typically > 10 Kohm). As a guide, subtract one standard load for the logic family from each driver's capability. This is necessary to ensure that contacts made during testing do not significantly alter the behaviour of the circuit under test.

8.5.2. Edge-sensitive inputs

Clock and other edge-sensitive inputs to the complete design should be buffered before use, for example using Schmitt trigger devices. This provides protection against unwanted operation of the components driven by the signals during testing, for example due to double-clocking caused by slowly rising edges from the ATE.

8.5.3. Test connections

Where components used in the design have dedicated test inputs or outputs these must be controllable from, or observable at, test points.

8.5.4. Unused inputs

Unused control or function-select inputs, test data inputs and controls must be tied to fixed logic levels via pull-up (for TTL), or pull-up/pull-down (for CMOS), or pull-down (for ECL) resistors, rather than through direct connections to power or ground. Where an input must be tied to zero (TTL) or one (ECL) this can be achieved through an arrangement such as that shown in Figure 8.15a, where the input to an inverter is pulled to the opposite logic level.

Groups of unused inputs can be connected to the same pull-up/pull-down network subject to the following requirements:

(1) Inputs controlling different functions (for example, load, clear, enable) in a component must not be tied together (Figure 8.15a).

(2) Functional data inputs for a component may be tied together.

(3) Groups of inputs controlling a single function may be tied together provided that the function can be enabled/disabled by changing the state of the tied network. (For example, where multiple chip select inputs are provided — Figure 8.15b).

(4) For 3-state enable inputs and other inputs that can force outputs to their high impedance state (for example, bus-request signals for microprocessors) it is preferred that inputs of different devices are *not* tied to common pull-up/pull-down networks. However, where this cannot be achieved, it is essential that 3-state enable inputs for devices that are otherwise connected together are not themselves tied together (Figure 8.15c).

These requirements allow the inputs to be used during testing if required. A particular benefit is to in-circuit testing, where the ability to apply standard library tests for component types can be compromised if the tester is unable to change the state of one or more inputs. Requirement (4) has a particular impact in in-circuit testing, since it allows component outputs to be placed in a high impedance state while the adjacent components are tested. This avoids the need for over-driving.

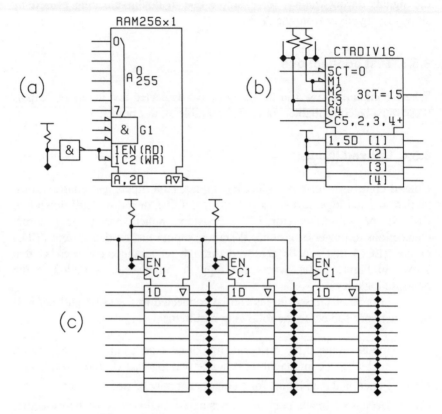

Figure 8.15 Requirements for 'unused' pins.

8.5.5. Wire-OR and wire-AND connections

The presence of wire-OR or wire-AND connections in a design complicates fault diagnosis since the ATE must determine which driver is injecting the faulty information onto the connection. Wire-OR and wire-AND connections should therefore be avoided if possible. Where they are used, care in assignment of driving gates into integrated circuit packages can considerably ease diagnosis problems (see Chapter 11).

8.5.6. 3-state connections and buses

Access

Those 3-state connections and buses that are not accessible at the design's functional interface (package pins, edge connector, and so on) must be connected to test points.

Termination

When none of the drivers is active, 3-state nodes enter a high-impedance state. This condition may occur either as part of the normal operation or due to faults in the various drivers.

When probed, fault-free nodes in the high-impedance state will cause indeterminate signals to be captured by the ATE hardware and may therefore appear to be faulty. This would significantly confuse the diagnostic process and must be avoided. All 3-state connections within a design must therefore satisfy one of the following requirements:

(1)   The connection must be driven at all times (except during transitions between drivers). Where this is not required for the basic design, additional drivers can be introduced to cover the 'dead' states (Figure 8.16 — option 1).

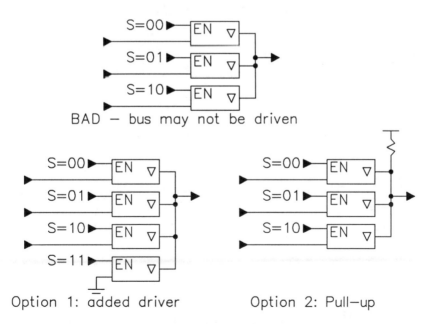

Figure 8.16 Avoid floating nodes.

(2)   The connection must be equipped with a pull-up or pull-down resistor of appropriate value that will force a known condition onto the bus in the absence of an active driver (Figure 8.16 — option 2).

Control of bus drivers

It must be possible to disable simultaneously *all* devices capable of driving a bus from a test point. It must also be possible for signals from test points to be used to determine which device is enabled onto the bus, in place of the built-in control circuitry.

These requirements allow the bus to be tested for shorts between wires without interference from bus drivers. They also allow the bus to be used as a means of testing the blocks of circuitry connected to it, independently of one another.

8.5.7. Feedback

The presence of global feedback paths (that is, feedback paths outside of stored-state devices) significantly increases the cost of test development and can complicate fault diagnosis.

In the former case, interference from feedback signals can make it difficult to propagate data needed for testing through the circuit. Data patterns may be corrupted by feedback signals, and paths may become blocked by unwanted feedback of control signals.

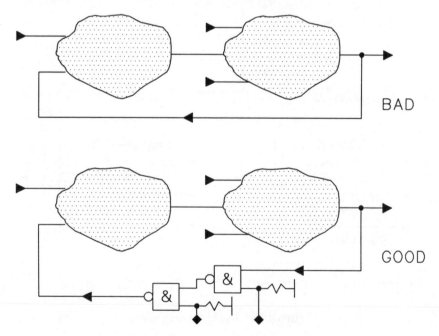

Figure 8.17   Break feedback paths.

In the latter case, diagnosis is impeded since faults on components in the feedback loop will appear at every node in the loop, and can only be distinguished by the time at which the faulty condition arises. Some test techniques (for example, signature analysis) consider the results of the test as a whole and are therefore unable to diagnose faults in feedback loops. Asynchronous feedback loops cause problems for all test techniques because faulty signals propagate almost instantaneously around the loop (see also Section 8.3.3).

To avoid these problems, facilities must be included in global feedback paths to allow feedback to be inhibited and/or for feedback signals to be replaced by data from test inputs at the appropriate stage of testing (Figure 8.17).

8.6. Controllability and observability improvement

In addition to the test access defined in the previous sections, the design should allow test access for control and/or observation at the following:

(1)   Data and address buses.

(2)   Read/write control lines.

(3)   Interrupt, hold, and halt lines.

(4)   Asynchronous controls.

(5)   Connections with fan-in or fan-out > 8 — for example, an output from an 8-input NAND gate or as in Figure 8.18.

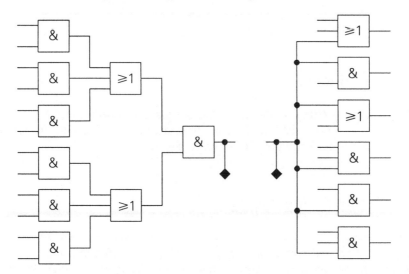

Figure 8.18   Placement of test points on connections with large fan-in or fan-out.

CHAPTER 9.
Supplying Power to the Product

9.1. Introduction

This chapter defines testability requirements for power supplies to the product and power distribution, including circuitry used to convert power voltage levels or to provide power line decoupling.

9.2. Safety during testing

The designer must ensure the safety of personnel involved in testing assembled printed circuit boards. In particular, the designer must ensure that test personnel are protected from (or made aware of) any safety hazards which might exist only during testing of the product, for example due to removal of casings, and so on.

9.3. Power inputs

9.3.1. Number of power supplies

The number of different power supplies fed to the product should be less than or equal to three. This limit is necessary to remain within the capabilities of programmable automatic test equipment.

9.3.2. AC power supplies

The use of AC power supplies to printed circuit boards should be avoided where possible.

9.3.3. Location of power pins

Power is supplied to the product through hard-wired connector pins or appropriate bed-of-nails access points. Therefore, to allow a text fixture to be used to connect more than one product to the ATE, it is advisable for standardized locations to be used for power pins. More information on this topic is contained in Chapter 10.

9.3.4. Power-on sequencing

The power supplies for a product must be able to be turned on in any order without affecting performance or initialization and without causing damage to the product.

9.3.5. Stabilization

The total test time for a product could be excessive if power supplies take too long to settle to their steady-state levels, resulting in higher than necessary test costs. The stabilization time for any input power supply must therefore be kept to a minimum, ideally less than one second (see Table 9.1).

Table 9.1 Stabilization time for power supplies.

Grade	Time to stabilize
Best	< 1 second
...	1 to 5 seconds
Worst	5 seconds to 1 minute
Prohibited	> 1 minute

Under no circumstances may a product power supply require in excess of one minute to achieve a stable level.

9.4. Decoupling

Power inputs to the following devices should be decoupled by the location of a capacitor as close to the component power pin(s) as possible:

❑   clocked devices (especially counters)
❑   devices forming asynchronous stored-state circuits
❑   devices feeding asynchronous inputs to stored-state circuits.

9.5. Conversion and validation

9.5.1. Regulators and converters

The product design must allow all voltage regulators and converters to be tested at full rated load current *before* power is applied to the rest of the product (thereby avoiding any possibility of damage due to faults in power circuitry). Where faults in built-in regulators or converters could cause damage to other components in a product, then the derived power supplies must be disconnected from the remainder of the product during testing.

9.5.2. Fuses and circuit breakers

All fuses and circuit breakers must be accessible during testing and capable of being replaced or reset without having to dismantle the product. (See also Chapter 11.)

9.6. Power-on resets

Where power-on reset circuitry is incorporated in a product, the design must also permit initialization by a logic input from the ATE. This ensures that the product can be reset to a known starting state at intervals during the test process, but without having to disconnect and re-apply power (see also Section 8.2).

CHAPTER 10.
Connector Selection and Layout

10.1. Introduction

There are two main ways of achieving satisfactory interconnection between the ATE and the **unit under test** (UUT) — through the product's normal functional connectors or through bed-of-nails probes. In either case, a test fixture must be constructed to convert between the format of the ATE's **zero insertion force** (ZIF) connector and the connector(s) used by the UUT (see Figure 10.1).

This chapter discusses how the product's connectors can be chosen and configured to reduce test costs. A key aim is to ensure that, where products use the same connector types, the layout of signals to connector pins is sufficiently standardized that a single ATE fixture can be used for all the products. If this cannot be achieved then a different test fixture will be needed for each product, resulting in increased test costs.

Design requirements aimed at reducing the cost of use of bed-of-nails test fixtures for printed circuit boards are contained in Chapter 11.

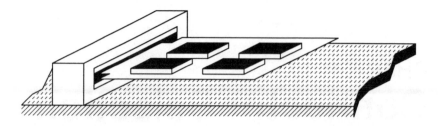

Figure 10.1   A view of the ATE pin electronics and test fixture.

10.2. Connector selection

10.2.1. Connector durability

The ATE fixture will be constructed using one or more connectors compatible with the interface(s) to the UUT. Since these connectors will experience a larger number of insertions/removals than the connectors on the individual UUTs, they must be chosen to give reliable operation in the test environment.

To ensure that ATE fixture reliability is not unnecessarily low, all UUT connectors must be chosen such that a high reliability companion connector is available that is specified to permit a minimum of 1000 insertions without significant degradation of its performance (for example, increased contact resistance).

10.2.2. Keyed connectors

Where a UUT connector is equipped with a locating key (for example, to prevent an incorrect board type being inserted in a system backplane slot), the key must be defeatable during testing to allow a common test fixture to be used.

10.3. Signal-to-pin mapping

10.3.1. Industry standards

Where an industry standard connector layout such as the VME-bus is used, it is recommended that the standard be followed without modification. This will permit the test engineer to reuse expertise previously gained in the operation of the interface.

10.3.2. Power and ground

Power and ground connections to the UUT will be hard-wired to appropriate connector pins. Changes in the location of the power connections will therefore usually require the construction of a new test fixture, and should be avoided.

A standard set of pins should be assigned to supply power and ground for all designs using the same connector type. Where a particular power supply connection is not used on a given design using the connector, then the pins assigned to that supply must not be reused for other purposes — either for alternative supply voltages or for signal connections.

10.3.3. Signal placement and grouping

For some test systems, certain facilities in the pin electronics (the electronics which drives signals into the UUT, or examines the UUT's response) cannot be set up on a pin-by-pin basis. The facilities must be uniformly used across groups of pins, typically across all sets of pin electronics contained on a single card in the ATE backplane (commonly 16 or 24 pins are driven or sensed per card — see Figure 10.1).

Other restrictions occur due to the differences in performance (and hence cost) of the pin electronics, which may mean that only a small number of the pins on a given test system may be capable of connection to an 'unusual' (for example, non-TTL compatible) signal on the UUT.

This section provides guidelines for the mapping of signals onto connector pins that will increase fixture commonality between board designs given that the ideal goal of total flexibility in the ATE will not be achieved.

Logic signals

The majority of ATE is equipped to handle TTL-compatible technology as the 'standard' option. Therefore, TTL-compatible logic signals can normally be freely allocated to connector pins.

For some lower-cost test systems, limitations may occur which limit the edge placement capabilities of the pin electronics. For example:

❒ only pins on the same card can change state at the same time;

❒ all pins on a card must have the same waveform characteristic (for example, return to 0/1/Z or surround by complement).

Designers should therefore keep signals that have particular timing requirements in fixed positions on connectors.

'Unusual' signals

'Unusual' signals such as:

❏   analogue signals

❏   high performance signals

❏   logic signals using unusual logic levels (for example, not TTL-compatible)

will need to be connected to ATE pin electronics with the appropriate capabilities. Such capabilities are likely to be confined to a small number of cards in the ATE system in order to reduce the cost of the equipment. The design of connectors for a range of cards using the same connector type should therefore ensure that such 'unusual' signals are wired to fixed connector pins.

External terminations

Typically the ATE pin electronics will provide for pull-up, pull-down, or a more complex programmable termination. These terminations will be needed at open-collector or balanced line driver outputs from the board, for example.

A common limitation here is that all terminations on a single ATE pin electronics card must (if used) be set to the same value, imposing a limitation on the design of the connector pin-out.

Problems arising from such ATE limitations can be avoided by ensuring that all signals that require termination use the same terminating network.

CHAPTER 11.
Printed Circuit Layout

11.1. Introduction

This chapter addresses the design and layout of printed circuit boards, including the grouping of logic elements (for example, NAND gates, and so on) into integrated circuit packages.

The principal objective is to ensure that the physical contact required between the ATE and the assembled board can be easily and reliably achieved. Two forms of access are typically required, in addition to access through the product's connectors as discussed in Chapter 10:

❐   through a bed-of-nails interface to connections within the circuit,
       either for in-circuit testing or to supplement access through the
       connectors during functional testing; and

❐   so that the ATE operator can manually probe inter-chip connections,
       and so on, as required during fault diagnosis.

An additional objective is to ease the task of diagnosing detected faults,
primarily by ensuring correct packaging of logic elements into integrated
circuit packages and easy identification of components.

   *The rules and guidelines in this chapter address only the test aspects
of printed circuit board layout. Designers should ensure that the
requirements placed on board layout to ensure compatibility with auto-
insertion equipment, and so on, are also met.*

11.2. Using this chapter

11.2.1. Board design requirements versus design-for-test

In Chapter 3 the impact of changes in integrated circuit and manufacturing
technology on design-for-test was discussed. Of particular note is the move
away from test techniques that depend on physical access into the core of the
board (for example, in-circuit testing) towards techniques that require access
primarily to the board's connectors (for example, ANSI/IEEE Std 1149.1).
These changes have a significant impact on the design-for-test requirements
for board layout.

11.2.2. In-circuit test requirements

For in-circuit testing, access is required to every component-to-component
connection on a printed circuit board. This is achieved using a bed-of-nails
interface fixture, as shown in Figure 11.1.

   To allow efficient and reliable use of a bed-of-nails fixture, it is
necessary to impose a number of 'accessibility' requirements on the layout of
a printed circuit board. These cover aspects such as:

❐   the ease of achieving a vacuum seal (typically, a vacuum is required
       to pull the board down onto the spring-loaded probes);

❐   the ease of contacting the probe target; and

❐   ensuring that components on the probed side(s) of the board do not
       interfere with probing.

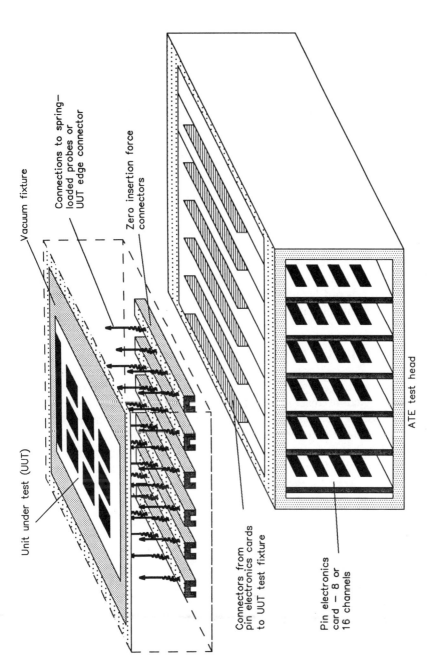

Vacuum fixture

Connections to spring—loaded probes or UUT edge connector

Zero insertion force connectors

Unit under test (UUT)

Connectors from pin electronics cards to UUT test fixture

Pin electronics card – 8 or 16 channels

ATE test head

Figure 11.1 Connecting to a board through a bed-of-nails.

11.2.3. Functional test requirements

For functional testing, a limited amount of access to internal connections using a bed-of-nails fixture may be required, for example to allow easier observation of internal circuit nodes or to gain access to test points added to improve the testability of the circuit design. In addition, physical access to the board will be required during fault diagnosis. This will typically involve the use of a hand-held 'guided' probe.

To ensure reliable manual probing of the board during diagnosis, aspects of board layout such as:

❐   the availability of probeable connections on the side(s) of the board accessible to the ATE operator during test; and

❐   the positioning and numbering of components

are of interest. After all, the aim is to test the assembled printed circuit board, not the competence of the ATE operator!

In addition, the manner in which logic devices (for example, NAND gates, flip-flops, and so on) are allocated into integrated circuit packages at the start of the board layout process can have a significant impact on the quality of the eventual diagnosis.

11.2.4. The impact of boundary-scan components

As was discussed in Chapter 7, by selecting components for use on a board that are designed to meet ANSI/IEEE Std 1149.1, the requirements for physical access to the board during testing are considerably reduced. Where a board is constructed entirely from such components, physical access will be required only to the product's connector(s) and to those connections within the board that form the test access path. The result is that components can be placed closer together, and that fewer artwork features need to be added to a board layout to ensure its testability.

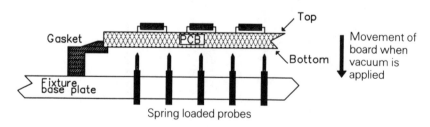

Figure 11.2 Board on ATE fixture.

11.2.5. Which rules should I follow?

Many of the rules and guidelines in this chapter apply to all boards, regardless of the test technique to be used. There are, however, some exceptions — notably the requirements for test access and component spacing. In these cases, the rules and guidelines apply only where in-circuit and/or functional testing are to be used.

11.3. Terminology

The terms 'top' and 'bottom' used in this chapter relate to the board as viewed by the operator when it is mounted on an ATE test fixture (Figure 11.2).

For dual-in-line, plated-through-hole board construction the component side is the top, while the solder side is the bottom. For other styles of board construction, for example where components are mounted on both sides of the board, the top side remains accessible to the ATE operator during test while the bottom side is contacted by the ATE, for example through a bed-of-nails interface. Note that, while fixtures are available that permit simultaneous access to both sides of a board, these are expensive. Also, in many cases the accuracy of probing for the top side of the board is less than for the bottom side, therefore some artwork features (for example, test access pad sizes) must be larger. The objective of this chapter is therefore to ensure that all printed circuit boards can be tested using a bed-of-nails fixture that contacts only one side of the board (the bottom), with the other side (the top) being available for manual guided probing.

11.4. Overall layout

11.4.1. Board shape

The preferred shape for a printed circuit board is rectangular with no cut-outs. Where cut-outs are used they must be surrounded by an area clear of all obstructions as discussed in Section 11.4.2.

11.4.2. Clearance around board edge, cut-outs, etc.

Bed-of-nails test fixtures are normally operated by a vacuum that draws the board down onto the spring-loaded nails. An amount of board area sufficient to accommodate an appropriate seal is therefore needed around the board edge, cut-outs, and any other features that could prevent a vacuum being established.

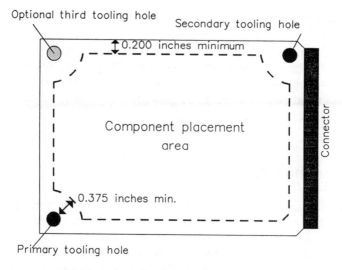

Figure 11.3 Tooling holes and clear area.

The clear area must be free of components, other obstructions, and via holes and should extend for at least 5.08 mm (0.200 inches) from the edge of the board, or cut-out. In the case of tooling holes, the clear area should extend for 9.52 mm (0.375 inches). This requirement is illustrated in Figure 11.3.

11.4.3. Locating holes

The accurate positioning of the board on the test fixture is critical if reliable connection is to be made to test lands and the edge connectors.

To allow accurate positioning, tooling holes with a diameter of 3.175, +0.100/-0.025 mm (0.125, +0.004/-0.001 inch) must be provided at two diagonally opposite corners of the board artwork as shown in Figure 11.3. If possible, a third tooling hole should be provided as marked. *All tooling holes must be un-plated*, because the accuracy with which the board can be positioned on the test fixture will be unacceptably reduced by the need to accommodate variations in plating thickness.

The positional tolerance of the tooling holes should be:

❏   +/-0.075 mm (+/-0.003 inches) between tooling hole centres
❏   +/-0.050 mm (+/-0.002 inches) between the tooling hole centres and other artwork features, such as test lands.

Note: The geometries defined in this chapter for other features (for example, test access point size) make the assumption that these requirements are met.

There must be a clear area of at least 9.52 mm (0.375 inches) annular radius around each tooling hole (see Section 11.4.2).

11.5. Interconnect, vias, etc.

11.5.1. Via holes should fill with solder during assembly

To ensure a reliable vacuum seal between the assembled printed circuit board and the test fixture, all via holes must be filled by the soldering process.

Via holes must therefore be of an appropriate diameter to ensure that they are normally filled during flow-soldering. Alternatively, provision must be made to ensure that the holes are filled following assembly.

There must also be sufficient clearance in any solder resist coating to allow via holes to be filled during soldering.

11.5.2. Use of mounting holes

To limit damage during repair of faulty boards, component mounting holes should not also be used to convey signals between layers of the board artwork.

11.6. Packaging of logic elements

When packaging logic elements into multi-element devices (for example, SN7400 NAND gates, or SN74374 latches) the following requirements must be satisfied where possible. These requirements should be satisfied in order from (1) — highest priority, to (3) — lowest priority.

(1)   Devices of the same type feeding onto a single wired junction or bus connection must be contained in the same package (Figure 11.4).

(2)   Gates, latches, and so on of the same type which are connected together must be located in the same package. *Note: This requirement is obligatory where an asynchronous feedback path flows through two or more elements of the same type.* (See Figure 11.5.)

(3)   Gates, latches, and so on of the same type which are in the same part of the design hierarchy must be located in the same package.

All these requirements, if followed, will ease the task of diagnosing a fault to a single replaceable part.

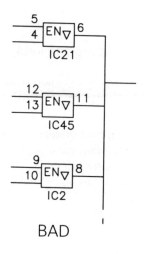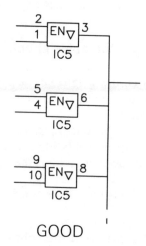

BAD

GOOD

If bus is faulty, then diagnosis
to failing driver is difficult.

Diagnosis of failing driver
is not required since all
drivers are in same pack.

Figure 11.4   Packaging of gates at wired junctions.

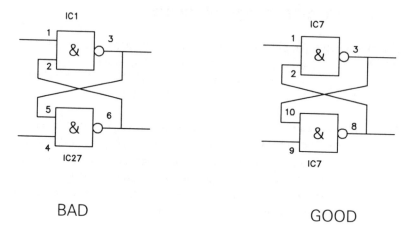

BAD

GOOD

Due to fast asynchronous feedback,
the ATE cannot distinguish faults
between the two gates. The only
possible repair action is to replace
both components.

Since both gates are in the same
package, diagnosis of the failing
gate is not needed.

Figure 11.5   Packaging of interconnected elements.

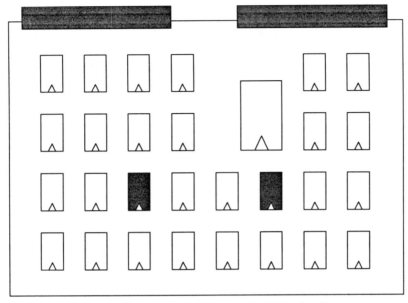

Shaded locations are for test—support circuitry.
Devices larger than one grid location spread into
neighbouring locations.

Figure 11.6 Regular board layout.

11.7. Component placement

11.7.1. Board layout

To maximize performance of the ATE operator or repair technician during diagnostic probing, the preferred overall layout of the board is for the main components (for example, integrated circuits) to be placed on a rectangular grid (see Figure 11.6). As discussed in Section 11.7.3, all components should be oriented in the same direction with pin-1 of each device being placed on a grid intersection. Components that are too large to fit within a single position on the grid can spread into neighbouring positions.

Note that Figure 11.6 shows a number of locations on the standard grid reserved for test support components. Since an amount of additional circuitry may be required to render the design testable according to the standards defined in this book, it is good practice to reserve space for the extra components from the outset — and to release it later if not required. Reserving a consistent set of locations for boards in a product range has benefits in terms of reduced cost of providing ATE interface fixtures.

11.7.2. Component mounting

The preferred style of assembly is single-sided component mounting on plated-through-hole printed circuit boards, using either through-hole or surface-mount components. Other styles of assembly, notably double-sided component mounting, significantly increase the cost of building ATE interface fixtures and complicate the ATE operator's task during diagnostic probing.

Where components must be mounted on both sides of an assembled board it is essential that the following factors are considered during component placement:

❏ Where practical, all integrated circuits mounted on the bottom side should meet IEEE Std 1149.1, since this removes the need for physical probing of their pins.

❏ Components on the bottom side may be in a vacuum during testing, which will inhibit cooling. These components must therefore be able to tolerate being in the vacuum for in excess of 15 minutes (to allow for fault diagnosis time).

❏ Components on the bottom side must not obstruct the operation of the bed-of-nails fixture. In particular, they must be clear of test lands (see Section 11.9) and their height must be less than 4.00 mm (0.160 inches) preferred, 9.00 mm (0.360 inches) absolute maximum.

11.7.3. Component orientation

The ease and accuracy with which an ATE or repair operator can locate a specified component pin for probing or examination can be significantly decreased if all components do not have the same orientation on the board. (Generally, a good operator can achieve around five probes per minute on a well laid out board.)

To allow maximum diagnostic throughput to be achieved, all components must have the same orientation, with their axes parallel to either the X or the Y axis of the board. Note also that for reliable flow soldering, all dual-in-line packages should have the same orientation.

Where it is necessary to adopt different component orientations, the following rules must be obeyed:

❏ all packages of the same style (for example, all dual-in-line packaged devices) should have the same orientation;

❏ package orientations should be at 90° intervals — that is, component axes must be parallel to the X or Y axis of the board; and

❏ the position of pin 1 and the flow of pin numbers (that is, clockwise or anti-clockwise) should be prominently marked beside each component.

11.7.4. Space between components

The rules contained in this sub-section must be followed unless neighbouring components conform to ANSI/IEEE Std 1149.1.

To allow for accurate and reliable contact between a guided probe, chip clip, and so on, and a component pin or printed circuit track, there must be at least 3.810 mm (0.150 inches) separation between component pins and adjacent components, measured perpendicular to the side of the component on which the pin is located. The space between adjacent sides of components which do not have connections must exceed 1.27 mm (0.050 inches).

These requirements are illustrated in Figure 11.7. Note that where sockets are used, the spacings are relative to the outer socket walls.

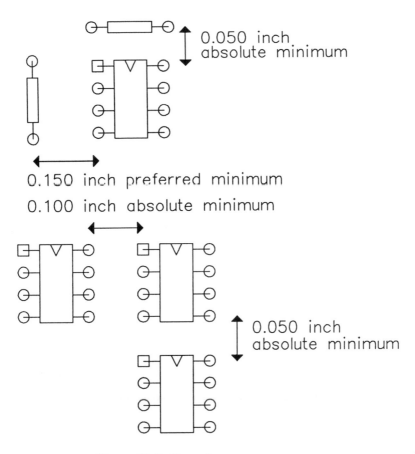

Figure 11.7   Space between components.

11.7.5. Rules for specific component types

Adjustable components

Adjustable components must be placed on the top side of the board. All adjustable components must be mounted so that they can be adjusted with a screwdriver or other instrument held perpendicular to the printed circuit board. The direction of adjustment should be the same for all components (for example, clockwise/anticlockwise, left-to-right, top-to-bottom, and so on).

Select-on-test components

Select-on-test components must be placed on the top side of the board. The mounting arrangement must be compatible with the number of insertions and removals anticipated during the product's life.

Indicators

Light-emitting diodes and other indicators must be placed so that they are visible when the board is mounted on the ATE.

Fuses and circuit breakers

All fuses and circuit breakers must be visible during testing, and capable of being reset or replaced without any dismantling.

Links and switches

Links, switches, and so on, must be placed on the top side of the board and be capable of being changed or operated without any dismantling.

11.8. Test access provision

11.8.1. Test access requirements

The following connections must be capable of being accessed through a bed-of-nails fixture:

(1)   all input and output connections of the board;

(2)   test access points included in the circuit design to improve its testability (for example, as defined in Chapter 8);

(3)   signals that connect to the TCK, TDI, TDO, TMS, and TRST* pins of components that conform to ANSI/IEEE Std 1149.1;

(4)   signals where one or more connected input and output pins is not a digital pin provided with boundary-scan capability to ANSI/IEEE Std 1149.1; and

(5)   each power connection (+5 V, ground, and so on) to a component whose signal pins are to be probed as a result of (4) above.

For a board populated exclusively with components that conform to ANSI/IEEE Std 1149.1, only the limited number of test access points defined by (1), (2), and (3) above is required. This limited set of test access points would also be sufficient if it could be *guaranteed* that in-circuit testing will not be used during the life of the product.

Note:

❐   Test access points must be provided for all otherwise un-used component pins to permit testing for solder shorts, and so on.

❐   Where jumpers, and so on, are used to complete a connection between two or more printed circuit tracks, each track is considered to be a separate interconnection and must be provided with its own test access point. Each signal connected to a jumper must be considered as an input or output connection of the board — category (1) above.

All test access points must satisfy the requirements on spacing, and so on, contained in Section 11.9 and be placed on the bottom side of the board.

11.8.2. Probe target provision

Prior to the addition of dedicated test lands to the printed circuit design, the following design features may be considered as locations for test access points:

(1)   via hole pads of diameter > 1.575 mm (0.062 inches) that are filled by solder during the assembly process and are not coated with solder resist, dry film, and so on;

(2)   leads of through-hole mounted **dual-in-line package** (DIP) components;

(3)   leads of through-hole mounted **single-in-line package** (SIP) components;

(4)   leads of through-hole mounted resistors, capacitors, diodes, and so on; and

(5)   plated connector tabs.

Where test access is not possible to one of the above design features, dedicated test lands designed according to the requirements of Section 11.9.2

must added to the printed circuit design. Optionally, dedicated test lands may be provided under other circumstances.

Where there are several acceptable probe targets for a single signal, the priority of selection followed during the fixture manufacturing process is (in descending order):

(1)   dedicated test lands;

(2)   via hole pads;

(3)   leads of through-hole mounted DIP components;

(4)   leads of through-hole mounted SIP components;

(5)   leads of through-hole mounted resistors, capacitors, diodes, and so on; and

(6)   plated connector tabs.

Note that access to the pins of surface-mounted components is not permitted due to the possibility of pressure from the bed-of-nails probes masking open circuit joints. *Dedicated test lands are therefore required for all interconnections made exclusively between surface-mount devices unless electronic access is possible to all component pins connected to the network.*

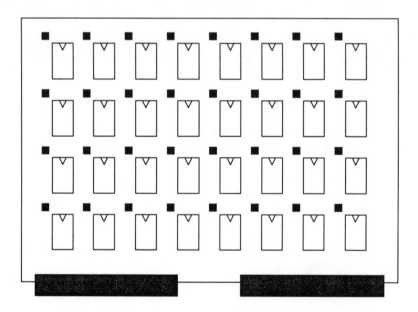

Figure 11.8   Regular array of test lands.

11.8.3. Standardized test point placement

Where only a limited number of test access points is required (that is, where all components conform to ANSI/IEEE Std 1149.1 or it can be guaranteed that in-circuit testing will not be used), it is possible to reduce the number of test fixtures required across a range of board designs. Examples would include cases where a board is one of a set forming a complete product, or where a widely used equipment practice is to be used (for example, Eurocard). In such cases, ATE interface fixtures can be reused if test access is achieved through dedicated test lands placed in standard locations. The benefit is substantially reduced cost of testing the product in production or repair.

Examples of standardized test land placements which could be adopted are shown in Figures 11.8 and 11.9. In the example in Figure 11.8, one test land is placed adjacent to each chip on the board.

Figure 11.9 shows clustered arrays of test lands surrounding board locations reserved for test support chips (for example, as described in Chapter 5). The locations shown are chosen to minimize the distance from the connection to be probed to a test land in the array. The locations reserved for test support chips may be released for other applications if not required for test purposes.

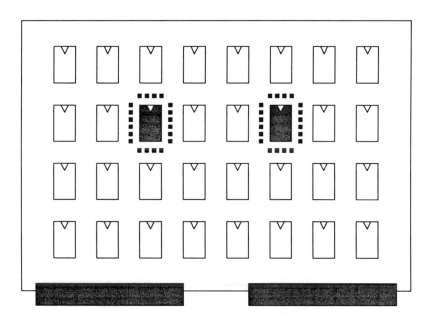

Figure 11.9 Clustered test land arrays.

11.9. Test access point design

This section details design requirements for test access points.

11.9.1. Placement

All test access points must be on the bottom side of the board. Where interconnections would not otherwise be accessible on the bottom side of the board, through-connections to dedicated test lands must be provided. A study of typical board designs (Bullock, 1987) showed that, in practice, a board will only contain a small number of such connections.

11.9.2. Dedicated test lands

Shape and size

Dedicated test lands may be square or circular. The square shape is preferred since this clearly identifies features of the printed circuit artwork which cannot be moved (for example, during engineering changes) without requiring modification of the test fixture.

The length of each side or the diameter (as appropriate) should be greater than 1.5 mm (0.060 inches) wherever possible, and must be greater than 1.0 mm (0.040 inches) in all cases.

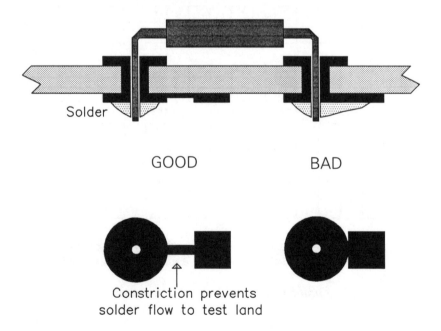

Figure 11.10 Separation of test lands from mounting holes, and so on.

Placement

Dedicated test lands must be separate from component mounting holes or pads. This avoids damage to bed-of-nails test fixtures due to attempts to contact uneven surfaces (for example, the solder meniscus) — for example, where the probe may slide off the solder surface and bend.

The provision of a constriction between the mounting hole/pad and the test land inhibits an excessive build up of solder on the test land during soldering (see Figure 11.10). Note that, while Figure 11.10 shows a through-hole mounted component, it is equally important that this rule is followed for surface-mounted devices.

11.9.3. Surface

The most reliable contact between a spring-loaded probe and an assembled board is achieved when the feature to be probed is coated with solder. This is because the softness of the solder allows the oxides that build up on its surface to be pierced by the probe.

The assembly process should therefore ensure that all artwork features that will be used as test access points are coated in solder (for example, via wave soldering, or by printing solder paste prior to component mounting and reflow).

11.9.4. Spacing

The spacing between test access points (for example, as listed in Section 11.8.2) must exceed 2.54 mm (0.100 inches) where possible, but may reduce to 1.27 mm (0.050 inches) where absolutely necessary. The objective is to minimize the number of probes on a 0.050 inch pitch since these probes can be unreliable in operation.

Note that it is not permitted to contact the leads of surface-mount components directly. Dedicated test lands must therefore be provided for all interconnections made exclusively between surface-mount package pins. Where package pins are spaced closer together than 2.54 mm (0.100 inches), track configurations such as illustrated in Figure 11.11 can be used to allow the minimum test access point spacing to be maintained.

11.9.5. Clearance around test access points

The space between the centre of a test access point and edges of any adjacent components mounted on the bottom side of the board must exceed 1.50 mm (0.060 inches) for all components, and 5.00 mm (0.200 inches)

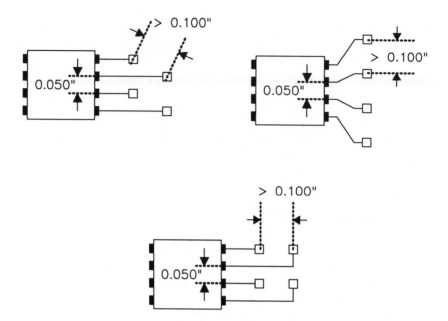

Figure 11.11   Test land placement for devices with pin spacings less than 0.100".

where the height of the component exceeds 4.00 mm (0.160 inches). These requirements are illustrated in Figure 11.12. Note also the maximum height for components on the bottom side of the board defined in Section 11.7.2.

The requirement for a component-free area around the test access points ensures that there is no interference between the spring-loaded probes on the fixture and the assembled board.

The requirement concerning tall components is caused by the limited space between the bed-of-nails fixture base plate and the bottom surface of the printed circuit board. To accommodate large components on the bottom side of the board the probe base plate needs to be shaped, perhaps by inclusion of a cut-out.

11.10. Labelling

11.10.1. Text size

All text on the printed circuit board should exceed 1.5 mm (0.060 inches) in height.

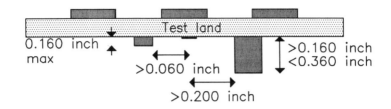

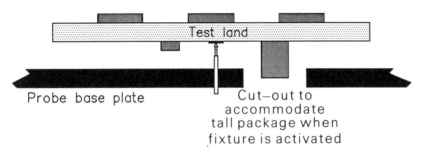

Figure 11.12 Clearance between test access points and components.

11.10.2. Text orientation

All text, including text on integrated circuit packages, must be at 0 degree (horizontal) or 90 degrees when viewed by the ATE operator. (See Figure 11.13.)

11.10.3. Board identification

The type, revision level, and serial number must be clearly marked on the top side of the printed circuit board.

Bar-code labels, where used, must also be on the top side of the board. (Note that the bar-code label may be the largest single item on a surface-mount board. The Texas Instruments SCOPE Diary component (Texas Instruments, 1990) allows board identification data such as revision and repair data to be held in an IC, rather than through a label. This component uses the ANSI/IEEE Std 1149.1 Test Access Port.)

11.10.4. Component identification

Unless all components are placed on a rectangular grid, all component identities must be marked on the printed circuit board adjacent to the component locations and must remain visible once the components are in place.

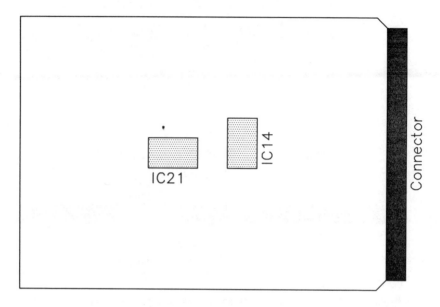

Figure 11.13 Text orientation.

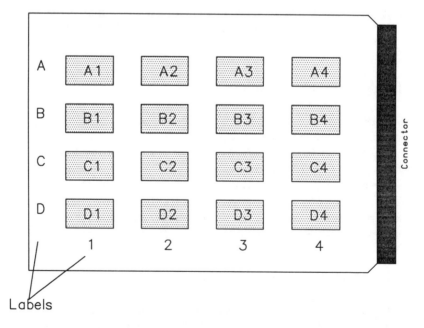

Figure 11.14 Labelling when grid layout used.

If components are placed on a rectangular grid, then component identities can be indicated by labelling the rows and columns as shown on Figure 11.14. Alternatively components may be numbered sequentially left-to-right, top-to-bottom, starting at the top-left corner of the board (see Figure 11.15).

11.11. Construction

11.11.1. Multi-board construction

Multi-board construction (for example, mother/daughter board) should not be used if at all possible, with the exception of the use of hybrids.

Where daughter boards (hybrids) are to be mounted on a mother board, access must be provided to all components and interconnections of the combined assembly without dismantling, through provision of test lands on the daughter boards where necessary. Particular attention must be paid to the ability to access test data and control signals through contacts to the mother board (see Section 11.9).

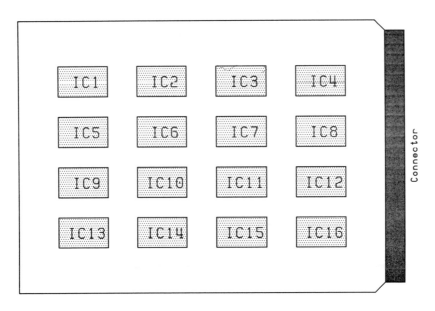

Figure 11.15 Sequential labelling of components.

11.11.2. Coatings

Where coatings are applied it is vital that points to which test access may be required are kept clear. Such points include:

❐   test lands on the printed circuit board

❐   via holes (to meet the requirement for vias to be filled by solder — Section 11.5.1).

❐   component leads

Coatings other than solder resist masks (for example, conformal coatings) should be avoided if at all possible.

11.11.3. Covers, and so on

Where all or part of a design is to be contained in a cover (for example for EMI screening) then the design must allow either:

❐   easy removal of the cover during testing, without affecting the operation or electrical performance of the circuit (EMC performance may, however, change); or

❐   test, diagnosis, and adjustment (where required) with the cover in place — that is, all faults in the circuit contained within the cover must be diagnosable with the cover on, and adjustments must also be possible with the cover in place.

References

Bullock M. (1987). Designing SMT boards for in-circuit testability. In *Proc. IEEE International Test Conference*, Washington D.C., 606-13.

Texas Instruments (1990). Scope DIARY data sheet.

CHAPTER 12.
Documentation

12.1. Introduction

This chapter defines the documentation required before detailed test development can commence. This documentation must be produced during the product development process.

12.2. Documentation required for test

The documentation generated by the product development team must include the following:

☐   *Bill of materials,* showing the types and (where appropriate) vendors of all components needed to build the product.

☐   *Detailed schematics.* Ideally, these should be provided as part of a copy of the complete workstation database for the design. If hierarchic schematics are not available, then the detailed schematics should be accompanied by appropriate block schematics.

❐ *Board layout data*, preferably in workstation format to permit extraction of the positional data needed to construct test fixtures.

❐ *Description of test-support features* built into the product (e.g., location of test lands, self-test, loopback, etc.).

❐ *Programming data for ROMs, PLDs, etc.* supplied as machine-readable files (for example, in JEDEC format).

❐ *Data for custom ICs*, including schematics, data sheets, test waveforms, and so on.

❐ *Set-up procedures* for adjust-on-test, select-on-test, switches, and similar components.

❐ *Performance specifications* showing maximum and minimum operating speeds, power consumption, and so on indicating especially those parameters that are critical to the successful operation of the product.

❐ *Timing relationships* between input/output signals, for example as would be found in a component data book, to show which signals precede others, set-up and hold times, and so on.

❐ *Functional test waveforms* for the completed design, to form the basis of any functional tests required for production or maintenance (for example, the waveforms created on the engineering workstation during simulation of the design). Note, waveforms should include both the inputs applied *and* the expected outputs (for example, as in a simulation log file). Note also that any test waveforms which must be applied in order to guarantee that the product meets relevant standards (IEEE, CCITT, Ethernet, and so on) must also be supplied.

APPENDIX.
Testability Checklists

A.1   Introduction

A.1. Introduction

This Appendix contains a set of testability checklists for use at appropriate stages in the development of a loaded board design.

General information on the use of these checklists is contained in Chapter 4. The checklists provide a reference to the section of this book where further information can be found on each specific topic.

Self-testing products

(1) Product identity

Product		Version	

(2) Rules. The product *must* meet these requirements.

Item	Rule description	Ref.	Yes	No	N/A
1	Self-test is triggered by: (a) power-up (b) user-interface command (c) electronic interface command	6.3			
2	Test results are indicated at: (a) user-interface (b) electronic interface	6.4			
3	No hazardous data is applied to the product interfaces during self-test	6.5			

Note: a negative response must be justified on an attached sheet

(3) Guidelines. Meet these requirements where possible.

Item	Guideline description	Ref.	%	N/A
1	Kernel is separated from other circuitry during kernel test	6.2		
2	Component-specific tests are applied	6.6		
3	A dedicated self-test ROM is provided	6.2.1		

Note: a response of <75% should be justified on an attached sheet

(4) Sign-off

	Role	Signature	Name	Date
1	Designer			
2	Test engineer			
3	Project manager			

Component selection

(1) Product identity

Product		Version	

(2) Rules. The product *must* meet these requirements.

Item	Rule description	Ref.	Yes	No	N/A
1	All device-specific testability requirements have been implemented	7.2.3			

Note: a negative response must be justified on an attached sheet

(3) Guidelines. Meet these requirements where possible.

Item	Guideline description	Ref.	%	N/A
1	Components are in the approved components list for the target manufacturer	7.2.1		
2	Simulation models are available for component used	7.2.1		
3	ICT test data is available for components used	7.2.1		
4	Components used contribute to the 'buy testable' policy	7.2.2		

Note: a response of <75% should be justified on an attached sheet

(4) Sign-off

	Role	Signature	Name	Date
1	Designer			
2	Test engineer			
3	Project manager			

Programmable device design

(1) Product identity

Product		*Version*	

(2) Rules. The product *must* meet these requirements.

Item	*Rule description*	*Ref.*	*Yes*	*No*	*N/A*
1	Device can be initialized	7.3.1			
2	All device outputs can be set to high impedance from a device input	7.3.2			

Note: a negative response must be justified on an attached sheet

(3) Guidelines. Meet these requirements where possible.

Item	*Guideline description*	*Ref.*	*%*	*N/A*
1	Device is synchronous	7.3.3		

Note: a response of <75% should be justified on an attached sheet

(4) Sign-off

	Role	*Signature*	*Name*	*Date*
1	Designer			
2	Test engineer			
3	Project manager			

ASIC design

(1) Product identity

Product		Version	

(2) Rules. The product *must* meet these requirements.

Item	Rule description	Ref.	Yes	No	N/A
1	Device can be initialized	7.4.1			
2	All device outputs can be set to high impedance from a device input	7.4.2			
3	Test programme exists that detects more than 95% of target faults	7.4.3			

Note: a negative response must be justified on an attached sheet

(3) Guidelines. Meet these requirements where possible.

Item	Guideline description	Ref.	%	N/A
1	Chip complies with ANSI/IEEE Std 1149.1-1990	7.4.2		
2	Scan or self-test techniques have been used where appropriate	7.4.4		

Note: a response of <75% should be justified on an attached sheet

(4) Sign-off

	Role	Signature	Name	Date
1	Designer			
2	Test engineer			
3	Project manager			

Circuit design

(1) Product identity

Product		Version	

(2) Rules. The product *must* meet these requirements.

Item	Rule description	Ref.	Yes	No	N/A
1	All stored-state devices can be initialized	8.2			
2	Prohibited initialization techniques have not been used	8.2.4			
3	Power-on resets are provided with an alternative logic control	8.2.5			
4	Unused initialization control pins are not tied together	8.2.5			
5	Analogue and digital circuits can be separated during test	8.3.1			
6	All redundant circuits can be tested separately	8.3.4			
7	RAM and ROM terminals are directly accessible	8.4.1 8.4.2			
8	ROM can be disabled and replaced by tester	8.4.2			
9	No monostables used in the design	8.4.3			
10	All built-in clocks can be observed and replaced by ATE signals during test	8.4.4			
11	No AOT or SOT components in design	8.4.7			
12	Fan-out margin provided to drive probes	8.5.1			
13	Test pins on devices are accessible	8.5.3			
14	Unused device pins are tied to fixed logic levels to DFT requirements	8.5.4			
15	Test points provided on buses and 3-state connections	8.5.6			
	More rules follow ...				

Note: a negative response must be justified on an attached sheet

Item	Rule description	Ref.	Yes	No	N/A
16	3-state connections cannot be high impedance in fault-free product	8.5.6			
17	ATE can determine bus driver during test	8.5.6			
18	Feedback paths can be disabled	8.5.7			

Note: a negative response must be justified on an attached sheet

(3) Guidelines. Meet these requirements where possible.

Item	Guideline description	Ref.	%	N/A
1	Initialization can be achieved within 20 clocks/patterns	8.2.2		
2	Static designs used in preference to dynamic	8.3.2		
3	Synchronous designs used in preference to asynchronous	8.3.3		
4	ROM contents can be made to give a consistent checksum	8.4.2		
5	Monostables meet all DFT requirements	8.4.3		
6	Long counters can be segmented for test	8.4.5		
7	Large parity trees, etc. can be segmented for test	8.4.6		
8	AOT and SOT components meet DFT requirements	8.4.7		
9	Handbag links, etc. used rather than switches	8.4.8		
10	Test points provided for LEDS and indicators	8.4.9		
11	Clock, etc. inputs enter via buffers	8.5.2		
12	No wire-OR, wire-AND connections	8.5.5		
13	Test access provided to key connections	8.6		

Note: a response of <75% should be justified on an attached sheet

(4) Sign-off

	Role	Signature	Name	Date
1	Designer			
2	Test engineer			
3	Project manager			

Power supply and distribution

(1) Product identity

Product		*Version*	

(2) Rules. The product *must* meet these requirements.

Item	Rule description	Ref.	Yes	No	N/A
1	Design offers no safety hazards for test personnel	9.2			
2	No more than 3 different power inputs	9.3.1			
3	Power supplied can be turned on in any order	9.3.4			
4	Stabilization time less than 1 minute	9.3.5			
5	Power inputs to specified devices are decoupled close to device power pins	9.4			
6	Regulators, converters, etc. can be tested without risk of damage to other components	9.5.1			

Note: a negative response must be justified on an attached sheet

(3) Guidelines. Meet these requirements where possible.

Item	Guideline description	Ref.	%	N/A
1	AC power supplies are not used	9.3.2		
2	Power pins are located in standard positions	9.3.3		
3	Stabilization time for power supplies (<1 sec = 100%, 1-5 sec = 75%, 5 sec-1 min = 50%)	9.3.5		

Note: a response of <75% should be justified on an attached sheet

(4) Sign-off

	Role	Signature	Name	Date
1	Designer			
2	Test engineer			
3	Project manager			

Connectors

(1) Product identity

Product		*Version*	

(2) Rules. The product *must* meet these requirements.

Item	*Rule description*	*Ref.*	*Yes*	*No*	*N/A*
1	A high-reliability companion connector is available for each connector used	10.2.1			
2	Connector keys are defeatable	10.2.2			

Note: a negative response must be justified on an attached sheet

(3) Guidelines. Meet these requirements where possible.

Item	*Guideline description*	*Ref.*	*%*	*N/A*
1	Relevant industry standard layout has been adhered to	10.3.1		
2	Non-standard signals are on consistent pins across product family	10.3.2		
3	Signal grouping requirements of target ATE have been met	10.3.3		

Note: a response of <75% should be justified on an attached sheet

(4) Sign-off

	Role	*Signature*	*Name*	*Date*
1	Designer			
2	Test engineer			
3	Project manager			

Printed circuit layout

(1) Product identity

Product		Version	

(2) Rules. The product *must* meet these requirements.

Item	Rule description	Ref.	Yes	No	N/A
1	5.08 mm clear area around board edge	11.4.2			
2	2 unplated tooling holes, in diagonally opposite corners	11.4.3			
3	9.5 mm clear area around tooling holes	11.4.3			
4	Via holes plated and clear of coatings	11.5.1			
5	Components on bottom side are < 9 mm tall	11.7.2			
6	All components oriented same way	11.7.3			
7	3.81 mm clear area from component pins	11.7.4			
8	1.27 mm clear area from component sides	11.7.4			
9	AOTs, SOTs, and indicators on top	11.7.5			
10	Fuses, circuit breakers, links, and switches accessible	11.7.5			
11	Test access point on bottom side for each node	11.9.1			
12	Test access points > 1.27 mm apart	11.9.4			
13	Test access points > 1.50 mm from components on bottom side	11.9.5			
14	Text at 0 or 90 degrees *More rules follow ...*	11.10.2			

Note: a negative response must be justified on an attached sheet

Item	Rule description	Ref.	Yes	No	N/A
15	Board identity, version, and serial on top	11.10.3			
16	Test access points clear of coatings	11.11.2			
17	Test and adjustment possible without removing covers	11.11.3			

Note: a negative response must be justified on an attached sheet

(3) Guidelines. Meet these requirements where possible.

Item	Guideline description	Ref.	%	N/A
1	Board is square with no cut-outs	11.4.1		
2	Mounting holes not used as vias	11.5.2		
3	Devices of same type driving same node are in same package	11.6		
4	Devices of same type that are connected together are in same package	11.6		
5	Devices of same type in same part of design are in same package	11.6		
6	Board layout on a regular grid	11.7.1		
7	Components mounted on one side of board	11.7.2		
8	Dedicated test lands on bottom, square	11.9.2		
9	Text height > 1.5 mm	11.10.1		
10	Component identities or numbering regular	11.10.4		
11	No daughter boards	11.11.1		
12	No coating other than solder resist	11.11.2		

Note: a response of <75% should be justified on an attached sheet

(4) Sign-off

	Role	Signature	Name	Date
1	Designer			
2	Test engineer			
3	Project manager			

Documentation

(1) Product identity

Product		Version	

(2) Documentation. The following documentation has been provided

Item	Item description	Yes	No	N/A
1	Bill of materials			
2	Detailed and block schematics			
3	Board layout data			
4	Documentation for test-support features			
5	Programming data for ROMs, PLDs, etc.			
6	Design data for ASICs			
7	Set-up procedures for AOTs, SOTs, switches, etc.			
8	Performance specifications			
9	Details of timing relationships between signals			
10	Functional test waveforms			

Note: a negative response must be justified on an attached sheet

(3) Sign-off

	Role	Signature	Name	Date
1	Designer			
2	Test engineer			
3	Project manager			

Index